THE COMPLETE GUIDE TO
BEEKEEPING

For my father Tom Evans, an ex-
beekeeper who offered to buy my
first hive; and for Bryony who was
the observer. . . .

THE COMPLETE GUIDE TO
BEEKEEPING

Jeremy Evans

In collaboration with
Sheila Berrett

UNWIN
HYMAN

First published in Great Britain by the Trade Division of Unwin
Hyman Limited, 1989.

UNWIN HYMAN LIMITED
15-17 Broadwick Street
London W1V 1FP

Allen & Unwin Australia Pty Ltd
8 Napier Street, North Sydney, NSW 2060, Australia

Allen & Unwin New Zealand Pty Ltd with the Port Nicholson Press
60 Cambridge Terrace, Wellington, New Zealand

British Library Cataloguing in Publication Data
Evans, Jeremy
   The complete guide to beekeeping.
   1. Bee-keeping-Manuals
   I. Title II. Berrett, Sheila
   638'.1

   ISBN 0-04-440374-7

The Complete Guide to Beekeeping was conceived, edited,
designed and produced by Julian Holland Book Design
and Packaging.

Designed by Julian Holland

Line illustrations by Martin Smillie

Photographers:
Jeremy Evans
Geoffrey Lawes
Margaret Nixon
National Beekeeping Unit, Luddington

Typeset by Input Typesetting and Microset Graphics

Printed and bound in Hong Kong.

Contents

Acknowledgements

Our thanks go to the following who have been particularly helpful with this book: Geoffrey Lawes who provided many of the more specialised photos, and read and criticised the final manuscript; Martin Smillie who drew the illustrations; Julian Holland who designed and produced the book; Arthur Chitty who dropped several pearls of beekeeping wisdom while down at his Stony Meadow apiary; Richard Strange who put the farmer's point of view; and Lesley who helped, aided and abetted throughout the project – I wouldn't have done it without her!

Introduction

This book was conceived in 1987. The year before I had determined to take up beekeeping which seemed a natural addition to the variety of birds and animals that my family were keeping at the time. I signed on for a series of beekeeping evening classes run by Sheila Berrett, and was immediately struck by her knowledge and extraordinary personality so 'why didn't we do a book together?', and by the end of that year it was underway.

Writing it and putting it all together has been an enjoyable experience – in particular working with Sheila and her husband Norman who was a willing model for the illustrations on many occasions. My thanks go to the many bees who were so co-operative with the photographs and put up with being prodded and examined on numerous occasions. The bees, Sheila, and I hope that you find the book as readable, useful and interesting as it was intended to be.

Jeremy Evans

Jeremy Evans

I first met Jeremy Evans when he joined my beekeeping class at Chichester College. Upon acquiring his first nucleus of bees he struggled his way through his first Beekeeping Year, encountering all the problems and difficulties which beset a newcomer to any subject. He then approached me regarding this book, asking if he could base it on the way in which I teach the subject. On thinking it through I felt that it really was a good idea – he would not gloss over matters which become irrelevant as one progresses, but are of paramount importance when you don't know and can't remember what to do next.

It gives me great pleasure to be associated with this very readable book, written in Jeremy's unusually gifted way, and which I hope will be used to help many potential beekeepers begin their long and happy beekeeping activities. It is not written in any way to prove that certain ways are absolutely right, but is merely a simple way for a beginner to start practising the craft. Once firmly established they may feel confident enough to start experimenting with other ideas and journeying further into the subject, at the same time obtaining many friends and great enjoyment.

Until that time I am sure that it will be very comforting for any beginner to have this book by their side.

Sheila Berrett

Sheila Berrett

Beekeeping – A 3 Year Cycle

Most of this book is based on 'The Magic Number 3', and it is therefore divided into the three year life of the queen bee – the first year the bee colony becomes established, the second it reaches maximum effectiveness, the third year it swarms. However there are limited hard and fast rules about beekeeping, not least because the vagaries of a climate such as the UK's will upset any supposed tables. A cold, wet spell at the wrong time can play havoc with a bee hive's honey production, which means that a beekeeper must be able to assess what his bees are doing and what they require next at any given time of the year.

This book will give you the basic ground rules of that knowledge, starting as a beginner, and 'qualifying' as an experienced beekeeper at the end of the three years. The problems you may encounter during the first year are dealt with in the troubleshooting section at the end of that year; after that we take a look at some more specialised applications and point you in the right directions to become more specialised still . . .

Apis Mellifera – the 'honeybee' – goes into action, searching for nectar to make honey and pollen to feed its young. A sophisticated, motivated, highly efficient worker and you will have around 50,000 of them in a hive! All you have to do is control and manage them efficiently, a skill which has been handed down over thousands of years of beekeeping.

The Rule of 3's

Almost everything about beekeeping runs in threes:

• There is a three year system to beekeeping. Buy a small 'nucleus' of bees in the spring of Year 1 and they will grow to form a nice colony. In their second year the colony will reach maturity, with fully drawn out frames for them to work on and lots of honey for you. In the third year they will want to swarm, and you will want to prevent them doing so.

This is an ideal sequence, but they may confound you in a number of ways – swarming in the first or second year, dying out one winter, or failing to reach maturity. As a beekeeper you have to spot the problems, and know how to remedy them.

•There are three 'castes' of honeybees – queens, workers and drones.

• The queen lasts three years as an effective egg-laying machine, though she can live to six years if the colony allows her too – a most unusual occurence.

• Three days for eggs.

The eggs are laid in the frames by the queen, and can be seen with the human eye. Each egg becomes a larva which soaks up the food that the worker bees supply it with, and after three days growth this diet will decide if it will grow into a queen or worker.

No one is sure of the exact diet of larvae during this time, but from day four a larva marked to become a queen (living in a 'queen cell') will be fed exclusively on Royal Jelly, while the worker larvae have to make do with pollen and a little nectar.

A further use of the number 3 in this respect is that worker bees first begin to secrete Royal Jelly, via the Hypo-pharyangeal glands in their heads, when they are three days old after emerging.

Knowing whether an egg is going to become a worker (the majority) or queen or drone (the minority) is important in second year beekeeping.

• Three ways of obtaining bees.

1 Buy a 'nucleus' – a small colony of around 10,000 bees headed by a young queen, clustered on four or more frames. The frames will contain new laid eggs, 'worker brood' (workers about to hatch out), food and pollen.

2 Catch a 'Prime Swarm' made up of approximately half an established colony which has left its old home headed by the old queen. Swarming is a natural impulse of bees designed to overcome overcrowding. The skilful beekeeper can control swarming to his own advantage.

3 Buy or be given a secondhand colony. You need to be very careful that you are getting disease-free bees and equipment.

• Three minute wait.

Beekeepers use a 'smoker' to help control bees if they become unsettled. If the bees sense a whiff of smoke they think their hive is on fire; they dive down into the combs to fill themselves up with honey which will act as emergency rations if they have to abandon the hive. This gorging on honey makes them soporific and less inclined to be aggressive; after the initial smoking the beekeeper waits three minutes for them to get to this stage.

• Three days before feeding a nucleus or swarm.

If you have just 'hived' a nucleus or swarm, it is usually best to wait three days before feeding those bees. The reason is that they are liable to be comparatively weak and too disorganised to protect themselves against 'robbing' by other bees if you feed them straightaway. They should have sufficient stores and no more, which will allow them to survive. A nucleus should be supplied complete with stores on the frames, while a swarm will have three days' stores in its bellies. So long as this is the case it's better to leave them enough time to sort themselves out,

post guards, and get themselves organised for foraging.

If you do need to feed them quickly, be very sure not to drip any sugar syrup round the hive; foreign bees will find it very quickly.

• Three day's honey in a swarm's honey sacs.

As mentioned above, bees about to swarm (or panicked by smoking) suck up enough honey to last them three days before they need to find more stores.

• Three feet or three miles if moving a hive.

Bees are like little robots controlled by a central computer – they always stick to a rigid flight pattern. If you move your hive a couple of miles from its original site, they will continue to fly back to that original site every time, hive or no hive, and will wait there patiently wondering what to do.

The rule of thumb is that you should only move a hive less than three feet, or more than three miles. If it is less than three feet the entrance can be aligned with the old entrance so they can find their way in; if it is more than three miles they will forget the old flight pattern, and establish a new flight pattern for the new site.

Hey grandad! Not so much has changed. The double wall hive is similar to a modern WBC; he's using a smoker; and the hive cover is a DIY version of a manipulation cloth. However nowadays we would prefer to wear just a little more protective clothing, just in case a few of those bees do decide to crawl up our legs!

The Rule of 3's certainly hasn't changed; nor is it likely to while beekeepers and honeybees continue to work together in harmony. Careful management and a little luck is what beekeeping is all about.

Beekeeping A–Z Glossary

A

Acarine
A mite which infests bee colonies.

Amalgamation
Hives can be amalgamated by putting one brood box on top of the other. A sheet of pin-prickcd newspaper is placed between. The bees eat through this, giving the colonies time to become acclimatised. The queen in the top brood box will usually be killed by the queen in the bottom box who will take over the amalgamated colony.

Artificial Swarming
This is an anti-swarming technique whereby you split the colony into two, preferably by early June. The beekeeper takes half the colony, mainly consisting of the queen and flying bees, and puts it in a new hive. The bees left behind will hopefully make their own new queen, and swarming will be averted.

B

Brace Comb
The workers sometimes build brace comb connecting frames, usually when they are incorrectly spaced. This can be scraped off by the beekeeper.

Bee Brush
A proprietry bee brush is indispensable for brushing bees off frames, etc. A feather will do the job, but not so well.

Bee Dance
A dance used by scout bees which tells the foragers what direction to take, and how far they will have to go.

Bee Escapes
Bee escapes are fitted into the feed holes of crown boards or glass quilts when required. Bees can go down through them, but can't get back up. A popular design is the Porter bee escape.

Bee Space
Bees need space to move between the frames, beneath the frames, and through a restricted entrance. Bee space for workers is about a quarter inch. If it is greater, they will tend to produce wax to fill the space with wild comb.

Brood
The brood area of frames contains the about-to-be-born bees in all stages from day old eggs to sealed brood with the developing bees covered by wax cappings.

Brood Box
A double depth box for extra deep frames to hold the main area of brood.

Brood Cells
The majority of brood cells are worker cells. Each colony should also have a number of slightly larger drone cells, recognisable by their domed cappings; and may have large acorn shaped queen cells if the workers choose to make them.

Brood Frames
Double depth frames for the brood box which accommodates the main laying area for the queen.

Brood And A Half
Brood box and super allowing the queen maximum laying space.

Burr Comb
Similar to brace comb – wax which is built up on top of the frames.

C

Candy
Bees can be fed a dense block of sugar candy with minimum water content during the winter.

Cappings
The workers cap over drawn-out cells with thin wax cappings. This is to protect the larvae as they grow into bees in the brood frames, and to protect stored honey which is capped once its water content has sufficiently evaporated. Whcn spinning-off honey, the cappings have to be cut off the face of the comb.

Casts
Casts are the swarms which follow the prime swarm, becoming proportionally smaller.

Cells
Each comb is a mass of cells in which the queen lays the eggs which are capped over and grow into bees called 'brood cells'; and the workers store honey and pollen for the colony.

Crown/Clearing Board
The crown or clearing board is used to cover the top set of frames. It has feed holes in it (usually two) which can be fitted with bee escapes to clear the bees down out of a honey super. The glass quilt is the same with glass panels and only one feed hole.

Cluster
The bees cluster together to keep sufficiently warm in winter. The cluster works its way over the frames, gradually consuming the winter stores.

Colony
At the height of the season in early summer the colony in a hive will number around 50,000 bees. This number drops substantially in the winter.

Colour
It is important to know the years in which queens and brood frames are introduced. A colour coding system is used.

Years ending in 0/5 Blue

Years ending in 1/6 White
Years ending in 2/7 Yellow
Years ending in 3/8 Red
Years ending in 4/9 Green

Combs
Bees build wax combs with perfectly formed cells for the queen to lay eggs in, and to keep their stores in. In hives these combs are built on wax foundation in frames supplied by the beekeeper.

Cut Comb
Pieces of honeycomb can be cut from unwired frames. Each frame should make five half pound honeycombs.

D

Dadant
An American design single wall hive.

Dividing Board
A solid brood shape frame, used to divide or limit the brood frames.

Drones
The male bees which are needed to mate with new queens. There are usually just a few hundred in a hive. Their other duties amount to no more than helping to keep the brood warm, so they have comparatively easy lives. They are recognisable by being much bigger than the workers.

Drawing-out
Worker bees have to build out the cells of foundation to a suitable depth by producing their own wax. This is called 'drawing-out'. Bees will always return to drawn-out frames, but need to be enticed onto new foundation.

E

Eggs
The queen lays up to 1,500 eggs per day. Each egg is laid in a brood cell, and finally becomes a new born bee some 21 days (if a worker) later.

Emergency Queen Cells
If a colony loses its queen, the workers can make emergency queen cells from eggs less than three days old.

Entrance Block
Hives have entrance blocks designed to completely close the entrance, or to limit it to prevent robbing.

F

Fast Feeder
A fast feeder is like a super filled with sugar syrup. It allows the bees to store the sugar syrup rapidly in the autumn when they are wintering down.

Feeder Bucket
Feeder buckets are used for feeding smaller amounts of sugar syrup.

Floor
The base of the hive which must be cleaned annually to prevent the build-up of disease.

Foragers
Adult workers leave the hive to become foragers, collecting nectar, pollen and water according to the requirements of the colony.

Foul Brood
A serious brood disease which usually requires the enforced destruction of the colony. The two varieties are labelled 'American' and 'European', though either can strike bees in both continents.

Frames
In the wild bees make their own wax combs. In a beehive, the beekeeper gives them wooden frames with sheets of wax foundation to work on in order to manage them efficiently.

Frame Feeder
The frame feeder is a hollow brood frame which can be filled with sugar syrup, and left in situ to feed a nuc or hived swarm. It also acts as a dividing board if you wish to cut the space of the brood box.

G

Glass Quilt
A glass crown board with the advantage that you can see what the bees beneath it are up to. Being prone to condensation, it should be replaced by a wooden crown board in winter.

H

Hive
A hive is quite literally a bee city, the beekeeper's home for bees which would normally live in the wild.

Hive Roof
The hive must have a roof to give full protection against the weather with a degree of insulation.

Hive Stand
A hive stand should be used to keep the hive clear of the ground. The WBC floor has a built-in stand.

Hive Tool
A indispensible steel tool which the beekeeper uses for freeing supers and frames which are propolised. Available in two principal versions, both with a flat wedge end.

Honey
The bees collect nectar from flowers, and convert it into honey which they store in the cells of the wax combs. The honey is food stores for the adult colony, but a large amount of it is stolen by the beekeeper.

Honey Extractor
Liquid honey is extracted by spinning the frames in a rotational extractor, driven by hand or electricity. Extractors are made in all sizes and at all prices. Most modern extractors are made in plastic.

Honey Flow
'Honey flow' really means an extensive nectar flow which keeps the bees going flat out, collecting and storing.

UK oil seed rape areas have an early honey flow around May. This can be followed by a 'June gap' when the nectar flow is insufficient for the bees; followed by the main honey flow when the nectar from flowers of summer and autumn plants starts rising.

L

Langstroth
An American design hive; probably the most popular in the western world.

Lifts
Lifts are the outer double walls of a WBC

M

Making a Nuc
A full grown colony can be split, with half the hive used to make a new nucleus of bees.

Mead
An alcoholic drink made with honey.

Metal Ends
Most beekeepers use removable metal ends to space their frames. Most are silver, but coloured ends are available which indicate which year brood frames are introduced to the hive.

Wide metal ends are also available for use with drawn-out frames in honey supers. The extra space between the frames gives the wax builders room to draw-out deeper cells for more honey storage. Wide ends must only be used above the queen excluder since they are unsuitable for brood. They must also never be used with new foundation – the bees will simply build wild comb in the extra space.

Mouse Excluder
A zinc strip which is attached to the hive entrance in the autumn to prevent mice entering in winter.

Manipulation Cloth
Canvas cloth on an alloy frame used to cover an open brood box or super when examining frames.

N

National
The most popular form of single wall hive used in the UK.

Nectar
Bees collect nectar which rises from flowers when the temperature is suitable. They convert this nectar into honey for their own stores.

Nuc Box
A small, half size brood box used to house a nucleus.

Nucleus
A young colony of bees covering a few frames.

P

Pepperpot Brood
Brood rejected by the workers due to some fault.

Plastic Ends
Coloured plastic ends are available to mark brood frames with the appropriate year.

Pollen
As well as nectar bees collect pollen from flowers, as their main source of brood food for developing bees. In doing so they help pollinate crops.

Propolis
A hard, resinous substance which bees collect to glue the hive together.

Q

Queen
The queen lays the eggs which allows the colony to survive. Nothing else is required of her.

Queen Cage
It is important to be able to recognise the queen. Once she is found, a round queen cage is used to isolate her on a comb so that she can be marked.

Queen Cells
The bees make queen cells to produce new queens, using larvae less than three days old. These may be swarm queen cells which are usually found on the bottom bars of the frames; supercedure queen cells; or emergency queen cells.

Queen Excluder
The queen excluder is a grid which prevents the queen getting to the honey supers.

Queen Paint
The queen should be marked with the colour of her first year, using a small blob of queen paint on her thorax just behind her head.

R

Requeening
Requeening is a technique designed to prevent bees swarming by introducing a new young queen, usually every second year.

Robber Bees
Bees go to the nearest, easiest source of honey. If foraging is in short supply, they will attempt to rob other hives.

Royal Jelly
A 'brood food' made by the bees which is initially fed to all larvae. From the third day it is only fed to larvae destined to become queens, while the workers receive a rather more ordinary brood food based on pollen with a little nectar.

S

Sections
Section honey is produced in square-sided frames with unwired foundation so that it can be eaten without spinning-off. Three sections are equivalent to one super frame.

Scout Bees
Scout bees tell the foragers where to go by means of the bee dance. They also direct a swarm to its new home.

Scrub Queen
A poor new born queen which will be rejected by the colony – a 'scrubber'!

Smith Hive
A Smith hive is a Scottish variation of a National.

Smoker
A smoker is indisensible if you need to subdue bees, but must be carefully used. When they sense smoke the bees dive down into the hive; they think their home is on fire and they have to fill up with honey.

Soft Cloth
Any kind of soft cloth which is large enough to cover a box of exposed frames is extremely useful when working a hive.

Solar Extractor
Used for extracting pure beeswax from all the residue left in a hive – cappings, brace comb, etc. Basically it is a double glazed box which relies on the heat of the sun to melt the pure beeswax.

Spring Clean
A hive should be thoroughly spring cleaned once the weather permits each spring. The brood frames are checked; new ones are introduced on a rotational system; and the main hive components are cleaned and sterilised using a blow torch.

Stings
Worker bees have a barbed sting. They can sting their natural enemies with impunity, but tough human skin will tear the sting out of their abdomens and kill them. The queen has a sting, but with no barbs; the drones have no sting.

Sugar Syrup
At various times bees need a helping hand with their food, not least because the beekeeper has stolen their stores of honey! They are fed large amounts of sugar syrup in preparation for the winter.

Super
The boxes which go on top of the brood boxes. These are used for the beekeeper's honey storage, separated from the brood by a queen excluder. A super is also used for a brood and a half.

Supersedure
Some colonies will automatically replace their old queen with a new young queen without any interference from the beekeepr. This is called 'super-cedure'. They won't swarm which is admirable, but if this trait continues they may begin to suffer from inbreeding.

Supersedure Cells
Queen cells which the bees make to produce a queen to supercede the old one.

Swarm
Bees have a natural inclination to swarm. They usually do this on a warm, still day around mid-day in spring or early summer. Half the colony leaves with the queen in a 'prime swarm'; the remainder stays behind with new born queens but may continue to swarm in smaller 'casts'.

Swarm Cells
In preparation for swarming, the workers make swarm cells to produce the new queens which will allow the colony to divide.

T

Travelling Box
A nuc box designed for transporting bees, fitted with a carrying handle and ventilation panels.

U

Uncapping Tray
Proprietary uncapping trays are available for removing wax cappings prior to spinning off honey. These can have a heated element to melt the wax.

V

Varroa
An infestation of a bee colony by mites. At the time of publication varroa was endemic in most countries, but had not reached the UK.

W

Wax
The word wax is derived from the Old English *weax*, as made by the honey-bee. They make it to build their combs, and when called upon to do so can produce a lot of it quickly – for instance a swarm can fill a skep with 'wild comb' in a week.

Beeswax is now superceded by synthetic waxes for many commercial uses due to its high cost. However it remains the peerless original.

Wax Foundations
Frames are fitted with flat sheets of wax foundation which the bees build out into their characteristic cells for laying and stores. You can buy worker foundation which is wired or unwired. You can also buy drone foundation with larger cells.

Wax Moth
Wax moth will lay its eggs in the wax of unused frames during the winter. The larvae will then systematically destroy the wax the following season.

WBC
A WBC is the pretty kind of cottage hive, named after William Brougham Carr.

Wintering Down
A hive must be carefully prepared for winter in the early autumn.

Worker Bees
The majority of the bees in a hive are female workers. Their duties range from working in the hive when they are young to going out and foraging when they get older.

The First Year: Starting

Your first year starts with the decision to take up beekeeping. You may have it enforced on you by inheriting or being given beekeeping equipment; or you may be starting from scratch. Whichever way you need a basic understanding of the commitment required before you become fully involved. Beekeeping is by no means a full time or expensive occupation, but at certain times of the year it does demand your interest and there is a certain amount of investment required before the honey starts flowing . . .

Learn from the experts. The chap bending down over the small 'nuc box' is judging new colonies of bees prior to a bee auction which is held annually at Brinsbury Agricultural College in Sussex. He describes each frame in detail, makes comments, and anyone could learn by watching and listening. Bee experts are invariably happy to pass on their knowledge; learn from them at classes, at association meetings, open days and other bee events.

When and How to Start

Prologue

• As a taster, try to visit the National Honey Show which takes place in October every year at Porchester Hall in London. It is organised by the BBKA (British Beekeepers' Association – see Appendices), and is packed with beekeeping equipment, exhibits and enthusiasts. You will find that beekeepers are generally always keen to talk about their craft and are extremely helpful to newcomers.

• Join your local beekeepers' associaton, most of which are BBKA affili-ated (see Appendices). The member-ship fee is low, and you should get a monthly newsletter and news of lectures – some also supply one of the maga-zines. Most local associations also have their own out-apiary, and have members' days for spring cleaning the hives at the beginning of the season, and spinning off the honey at the end of it. In the early autumn many organise local honey shows.

• Visit you local beekeeping supplier. They are usually enthusiasts who will discuss what you need to buy from them and when.

• Try the bee journals that are avail-able, and look through the manufac-turers' catalogues.

• Many local authorities and some agricultural colleges run beekeeping evening class courses which enable you to discuss plans and problems with an expert, and to meet other novice beekeepers.

Some courses start in October, but we would recommend starting in January. The first term should intro-duce you to basic theory; the second spring term should include practical lessons keeping pace with your own beekeeping through to the early summer months; and the third autumn term should help you to winter your bees down, and then prepare for the next year.

Preparation

If you are starting from scratch you will need to buy a nucleus of bees. The year's new queens start emerging in May, after which the small nucleus colonies build up rapidly ready for you to take delivery in late May/early June. Timing is as always 'weather permit-ting'. If the weather is fine a nucleus will expand rapidly, as the worker bees are able to get out on plenty of foraging trips to provision the hive and encourage the queen to lay as fast as she can; if the weather is foul, the bees will be more hive-bound and limited in their foraging, slowing the growth of their colony.

However before taking delivery of your bees there are three important things to do:

1 Choose the site for your hive.

2 Decide what type of hive you are going to use, and order in plenty of time. Beekeeping always has an early season rush, and if you order late you may find there are several weeks' delay, particularly if you want a 'ready made' hive.

3 Work out a rough timetable for the year ahead, and remember that like having a dog for Christmas beekeeping is a commitment that sometimes cannot be ignored. The first year is relatively undemanding, though you should be too interested to take no notice of your new colony; but from the second year onwards the bees will take more of your time if you want to maximise honey production, and you must be willing to work on the hive at the drop of a hat in the month of May.

Live bees for sale! A nucleus of young bees freshly bred for the new season gives you a starter colony which will expand rapidly in May/June time.

Beekeeping Equipment

Beekeeping equipment in such countries as the UK and USA is relatively sophisticated, but elsewhere they do not take such luxuries for granted. In China a beekeeper is unlikely to have any protective equipment, and will make do with a spoon as his only hive tool. However you need to be highly experienced to approach bees in such a manner, and the wealth of western equipment makes beekeeping far more accessible to novices.

To start you must have a minimum amount of protective clothing, most important of which are a combined hat and veil.

You will need a hive with a floor; brood box with frames for the queen to lay in; a super with frames for the workers to store honey in; a glass quilt or crown board to cover the top of super or brood box; and a roof to keep the rain off.

You will have to feed the colony to start with, so a small 1/4 gallon plastic feeder bucket is desirable at the start of the season, and later on you will need a larger 1 gallon feeding bucket or fast feeder for wintering down.

If the colony multiplies rapidly you may also need a queen excluder to separate the laying frames from the honey frames when and if you start harvesting your first honey.

Added to this the basic tools you will need are a smoker and a hive tool. A manipulation cloth, used as a temporary cover for the brood box or super, is also extremely useful when doing any work that involves lifting out frames; a second soft cloth is also necessary. A bee brush is another extra which you may find necessary, although a long bird's feather will work just as well.

If you are buying this equipment new, the minimum outlay is likely to be around £150; or spend twice that amount, and you can be very well equipped with the very best gear. You may be given some of the gear or buy secondhand (page 47) which will of course reduce your overall spending.

Looking further ahead, you will need more supers and the frames to fill them in the second year. At this stage you should also consider getting a second hive, which serves as a useful back-up if one colony becomes sadly depleted due to swarming or wintering badly.

THE SMOKER

A smoker consists of a container to hold a slow burning fuel; a bellows to make it burn faster and produce smoke when required; and a nozzle to direct the smoke at the bees. The classic style of smoker is copper bodied and will last a lifetime. You can buy the same type in tin plate which is cheaper, but will corrode after a few years. Larger, commerical size smokers using the same principle are also available (only in tin plate), designed for working on up to 20 hives at a time!

Another type of smoker has a clockwork driven fan in place of a bellows. You wind up the clockwork, and simply press a release button to activate the fan when you want smoke. This has the advantage of being a one-handed operation, and makes it considerably easier to keep the smoker alight. The same principle could no doubt be used with an electric motor.

Lighting a smoker

If you take care with lighting a smoker it will stay alight and puff sufficient smoke for at least 30 minutes, which is more than enough time to inspect two hives. Even if you are unlikely to need smoke it is always a wise precaution to have your smoker ready to hand and lit when doing any work on a hive that interferes with the bees.

Ideal fuel to get a smoker going is corrugated cardboard paper. This can be ripped off any old cardboard box, but in many cases will have been saturated with a fire retardant to prevent it burning. To get rid of this you have to soak it in water, and then dry it. Then, when you are ready to go to the hive, proceed as follows:

All smokers work on the same principle, with a bellows attached to the main fuel canister. The design is unfortunately very antiquated. A modern smoker which keeps going whenever you need it is sorely needed by the beekeeping community.

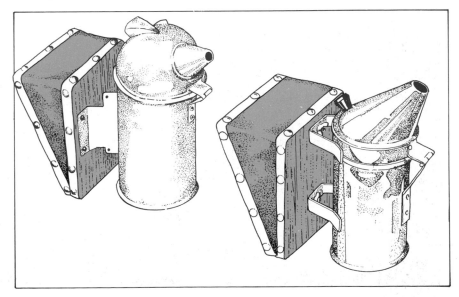

1 Roll the cardboard into a tube, trimmed to be big enough to fill most of the container with a little space round the outside and at the top.

2 Pull out some of the middle to form a wick, and set it alight. A cigarette lighter is handier than messing around with matches, particularly if you have to relight the smoker in the middle of working on a hive.

3 To get the cardboard well alight, wave it back and forth until the flames shoot out of one end.

4 Then invert it flame downwards, and place it in the container.

5 Fill the empty spaces with dry grass, wood shavings, or bits of hessian sacking if you can get it (the ideal fuel, but virtually unobtainable these days). Anything that will smoulder is a good fuel, but don't jam the container too full or you will smother the fire.

6 Give a few puffs on the bellows to make sure everything is well alight. Then when working on the hive take time to give the occasional puff (away from the hive unless you wish to smoke the bees) to ensure that your smoker keeps going. When you have finished with it, you can extinguish the fire by ramming fresh grass down the nozzle, and the unburnt fuel can be reused next time.

It you have problems getting your smoker to light, using half a firelighter may help. Take spare fuel (dry grass or wood shavings) to the hive with you in case you need to top it up, and when putting the smoker down remember that the copper body becomes very hot and will burn holes in a synthetic material such as a plastic feeder bucket.

The smoker is an important piece of beekeeping equipment, and we would recommend having a practice run with one before using it in earnest. Just light it up to let it burn and see how it behaves. So long as you buy the copper bodied version the only necessary upkeep is cleaning, and that only if you're houseproud!

The importance of the smoker

It is worth emphasising what the smoker is and is not for. It is used to sedate

the hive, but if used too enthusiastically may well make the bees become extra aggressive – just the result which you don't want!

In the pages which follow we indicate which operations are likely to require a little smoke, though there are no hard and fast rules as you always have to take bees as you find them. The basic technique is to first give a gentle puff of smoke it the entrance to make them

The smoker should be used cautiously, but be ready nonetheless. The hives are Nationals, piled with supers.

head for the honey and suck it up ready to abandon home – but don't smoke them too hard or you will drive them out of their minds.

You then wait for three minutes before taking the roof of the hive off. By then the bees will have been sedated

LIGHTING A SMOKER

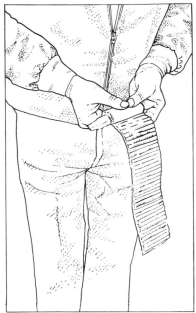

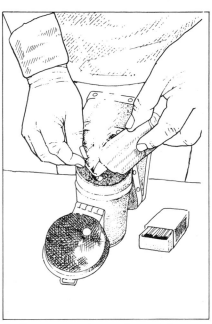

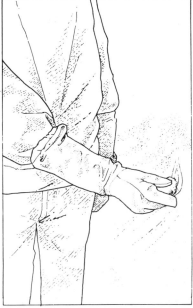

Roll up some cardboard. If it has been treated with a fire retardant, you should soak your cardboard and allow it to dry.

Pull out one end to make a wick. Light it in a draught free place. Your smoker should already contain some fuel.

Wave it back and forth until it is well alight, with flames shooting out of one end. A glove may be useful!

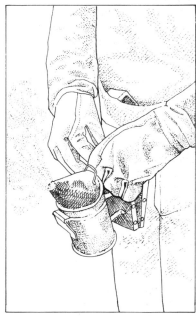

Carefully push the cardboard down into the smoker with the lighted end downwards. Have more wood shavings (or similar fuel) ready.

Cover the top of the cardboard with this fuel. Close the top of the smoker, and give a few pumps on the bellows to make sure it is going.

The smoker should stay alight for at least 30 minutes, and can be refuelled as necessary. Block the nozzle with grass to put it out.

by their large intake of honey. As you remove the glass quilt (or clearing board) which is covering the super or brood box give a few gentle puffs round the edges to clear your way, but no more or you will begin to antagonise them.

The bees should be be sedated as you work on the hive, but will be ready to go if the imaginary fire appears to be taking hold. If you oversmoke them you could drive out the whole colony, many of whose members will be gunning for you; if there is no more smoke they will assume the fire is no longer a problem and gradually revert to normal.

HIVE TOOLS

A purpose-designed hive tool is invaluable for the following reason. The bees like to glue everything together with 'propolis'. Thus the super is glued hard to the top of brood box, the glass quilt is glued hard to the top of the super, and the frames are glued hard to one another. You could look on it as a sort of advanced form of draught proofing.

There are two basic types of steel hive tool available with slightly different characteristics. Unfortunately a single tool which combines the best of both has not been invented.

The yellow hive tool

The yellow painted hive tool is wider and heavier, and is generally favoured by commerical beekeepers. One end is used to lever open things held together by propolis, and does so without too much damage to the surrounding wood; the other acts as an efficient scraper of unwanted propolis and other gunge. The lever end is the width of a frame, and if the box isn't full can be twisted between two frames to make the right space to insert a new frame.

The red hive tool

The red hive tool is slimmer, with one end incorporating a notch designed to lever each frame clear of its box and then be lifted out. The principle is not dissimilar from that of an old can or bottle opener. The other ends acts as a wedge to lever open propolis, but is narrower than on the yellow version, and is liable to inflict more damage on the surrounding wood.

The truth is that you are best off with both kinds of hive tool, and at some £5 each it is no great extravagance. Like the smoker the hive tool is an aid, but remember to use it in a calm, methodical fashion so as not to antagonise your bees. They dislike sudden movement or bangs, which are liable to result if you blunder ahead without carefully thinking out what you are going to do before starting to tackle a hive. Think before you act!

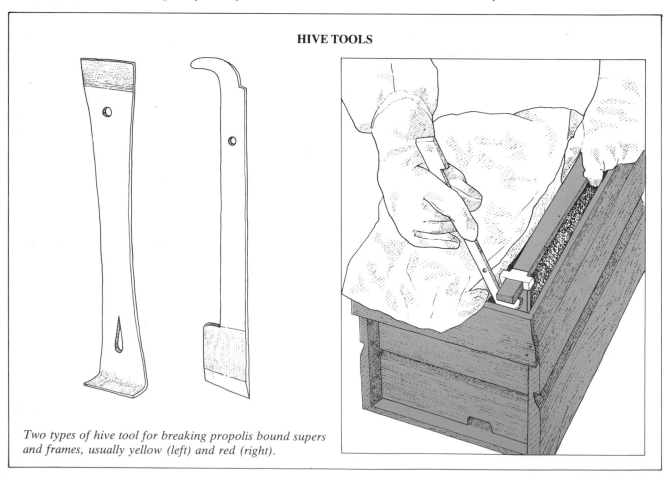

HIVE TOOLS

Two types of hive tool for breaking propolis bound supers and frames, usually yellow (left) and red (right).

22

The Hive

The hive is quite simply the bees' home. It is not the bees' natural home, but is the beekeeper's idea of how to keep the bees in a controlled environment where they can produce the maximum amount of honey for us humans to use. Bees would be quite as happy living in an old log or up under the eaves of your house, but you would not have much luck harvesting their honey or taking care of them through the winter.

There are many different designs of hive around the world, but the two you are most likely to see in the UK are the WBC (shown below) and the National (shown right) because the frames are interchangeable.

The WBC is the traditional style of country hive which you expect to see in the back gardens of thatched cottages. WBC stands for William Broughton Carr, the famous beekeeper who invented it at the turn of the century, and though the prettiest ones are painted white it is now more usual to see them in unpainted hardwood which will last a lifetime with very little upkeep.

The National works on the same principle as the WBC, but lacks its characteristics outer walls which are known as 'lifts' – lift them off and you will find a similar floor, brood box and supers. However be warned that brood box and supers are not interchangeable from a National to WBC or vice versa, so once you have opted for one type of hive you should stick with it. The frames are the same size, but while the WBC brood box and supers have 10 frames the National's have 11 to give the hive slightly more capacity. The Langstroth and Smith are other types of popular hives detailed overleaf, using different size components, though the principles and working them remains the same.

The exploded view of a National hive overleaf breaks it into its component parts:
- Blocks to raise it off the ground and prevent damp.
- A floor which is the base of the hive and contains the bees' front door entrance.
- A brood box, which is the main year round home for the bee colony.
- A super (super meaning 'on top of') which is the food or honey store for the bees, with a queen excluder to prevent the Queen laying eggs in it.
- A roof to keep the weather out and give the bees a nice, snug home.

Cutaway view of a WBC hive with brood box and honey supers. The queen excluder is always positioned between honey supers and brood, and extra supers can be added as the bees work through the frames. A brood box and four supers is as high as you're likely to go in a very good year,

Note how the sliding entrance blocks vary the openings. Each of the little cut-outs is an approximate bee space.

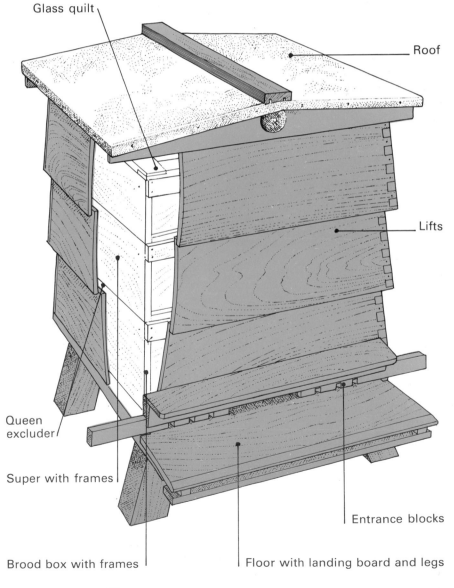

Glass quilt

Roof

Lifts

Queen excluder

Super with frames

Entrance blocks

Brood box with frames

Floor with landing board and legs

23

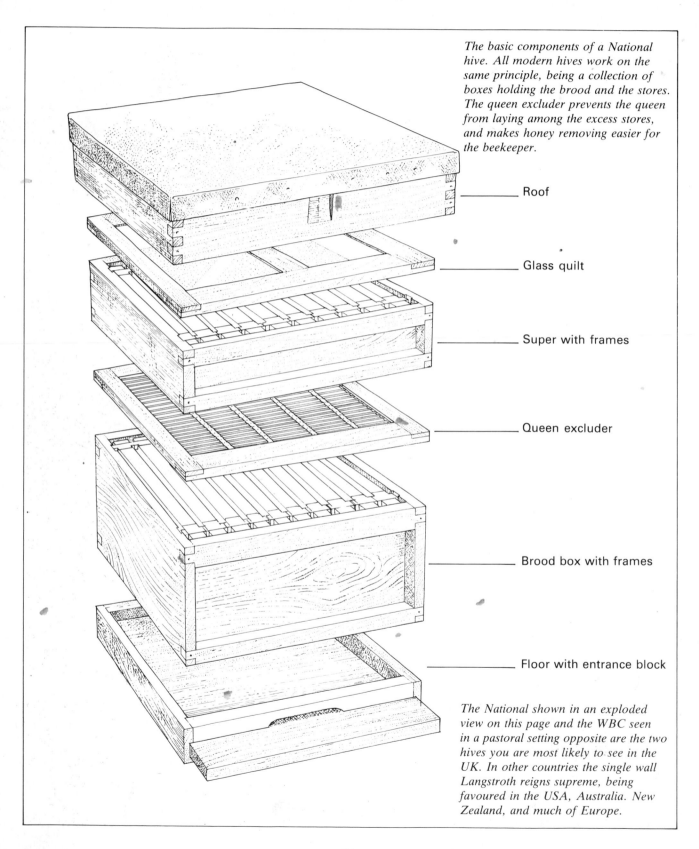

The basic components of a National hive. All modern hives work on the same principle, being a collection of boxes holding the brood and the stores. The queen excluder prevents the queen from laying among the excess stores, and makes honey removing easier for the beekeeper.

Roof

Glass quilt

Super with frames

Queen excluder

Brood box with frames

Floor with entrance block

The National shown in an exploded view on this page and the WBC seen in a pastoral setting opposite are the two hives you are most likely to see in the UK. In other countries the single wall Langstroth reigns supreme, being favoured in the USA, Australia. New Zealand, and much of Europe.

More Hives

While the National is the most popular hive in the UK, the American Langstroth is without doubt the most popular hive in the USA and worldwide.

The Langstroth looks like a National and works in exactly the same way with very similar components, and the main difference is that it is larger – (20″×16¼″ [51×41cm] as against the National which is 18⅛″ [46×46cm] square), with deeper boxes to increase the brood and honey area.

It is available in the same choice of pine or cedar, with prices that are only slightly more for a bigger hive. The Langstroth offers a brood area of 2,750 (17,750 sq cm) square inches against the 2,200 (14,200 sq cm) of the National.

Langstroth advantages

The larger brood area equals greater capacity which in principle means less hives and less work – one hive doesn't have to be worked on so often – for the same amount of honey.

Many commercial producers of honey use Langstroths for these reasons, and they work very well if there is a long enough honey season to maximise their capacity. In some countries even larger hives are used which require heavy lifting gear.

Langstroth disadvantages

For use in the UK we feel there are several disadvantages.

The Langstroth is totally noncompatible with the popular National and WBC, and since Langstroths are not widely used there is little secondhand gear available. Commercially they may seem preferable, but a National can be boosted to similar capacity by using the optional National 'commercial' or 'jumbo' brood box (as deep as a conventional brood and a half). All the National components remain compatible, and the hive can be cut back down to a more manageable size during the winter, or when the beekeeper can no longer handle the extra weight and size

of his hive through age or illness.

There is no way the Langstroth can be cut down to the size of a National, and while a National super is heavy at 30lb (13.60kg) full, a Langstroth super is even heavier. Furthermore the capacity of the Langstroth brood box is so large that a hobbyist beekeeper may well find that during a normal season the bees don't get round to filling the supers and he consequently has to raid the brood box – a more fraught procedure – to take an appreciable amount of honey off.

Smith Hive

The Smith hive is a Scottish hive which is popular north of the border. While it looks similar to a National its boxes are non-compatible and take short lugged frames. We would not recommened the Smith as a new buy, but it may be a good buy secondhand. National frames can be made to fit by the simple expedient of sawing off the lug ends.

Dadant Hive

Like a Langstroth, but even bigger with a massive 3,740 square inch (24,000 sq cm) brood area afforded by 11¼″ (28.50 cm) deep brood frames. This hive in only for serious commercial use.

Final Choice

The Langstroth is very successful abroad in countries where beekeeping is a male preserve and where young, strong men work hives commercially.

For the UK hobbyist a National is much more serviceable, is widely available new or secondhand, and can be increased to semi-commercial size. The WBC is more picturesque, but with its many parts is more fussy to handle and cannot be recommended to migratory beekeepers who want to move hives to take advantage of various crops.

The way things were. Before modern hives beekeepers kept their bees in skeps, made here by Dick Tutton.

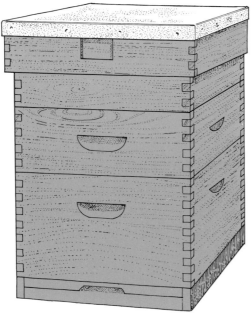

LANGSTROTH

SMITH

DADANT

Langstroth, Dadant and Smith are all single wall hives similar in design and working on the same principles as the National. All these single wall hives are more practical to work than the double wall WBC with its outer 'lifts', and are also less costly. They simply consist of floor, brood box, supers and roof, and can easily be made by the DIY carpenter.

The Langstroth is very popular in the USA and elsewhere since its large capacity can give very good returns. However full supers and brood box are heavier to lift than a smaller National, Smith or WBC which are therefore to be preferred by those who doubt their weight lifting abilities, particularly when dealing with angry bees!

Hive Components

It is worth examining the components of a hive in detail, starting from the bottom.

Bricks or hive stand

The hive must be mounted on bricks or a hive stand to keep it clear of the ground. Damp kills bees, and it also rots wood.

Floor

The floor is basically a flat wooden surface with a slot cut out for the bees' front entrance. The size of this entrance can be adjusted or completely closed by means of an entrance block (National) or slider (WBC). On a WBC the floor also doubles as the stand for the hive.

Entrance block

The entrance block on a modern WBC is a two-part slider which can be adjusted from fully open to a few 'bee space' holes, or completely closed. A bee space is $^3/_{16}$–$^1/_4''$ (4.7–6.3 mm) square to allow one worker to pass in and out – if the hole is any larger the bees will follow their natural instinct and block it with propolis.

The entrance block on a National is a more basic square ended piece of wood. One side has no cut-out to completely close the entrance; the other has a large cutout to allow the bees a limited amount of access.

Being able to control the size of the entrance is primarily important when the bees have to defend the hive against robbing. If the hive is very strong the bees can take care of themselves and will fare better with an unobstructed entrance; but in the early days of a nucleus or after a long, hard winter the colony is weak, and will find it much easier to protect itself if the size of the entrance is kept small. It's a bit like Erroll Flynn fighting off a dozen brigands from the top of the stairs.

You should also reduce the entrance to a minimum when feeding your colony in preparation for the winter, but when

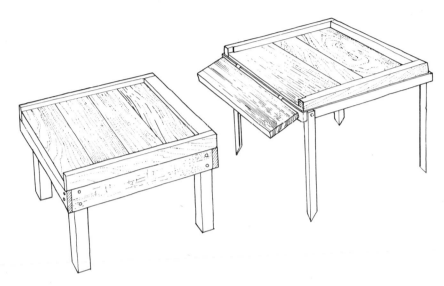

Above: Nationals and other single wall hives need a stand to keep them off the damp ground. Some beekepers use a milk bottle crate as a cheap option!

Below: The WBC floor has legs attached and therefore doesn't need a stand. It is wise to protect the undersides with roofing felt.

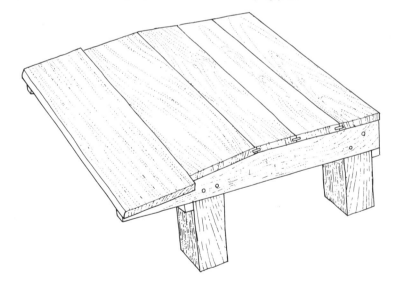

winter comes the entrance should be completely opened to keep the hive well ventilated. The danger is then likely to come from mice looking for a home: to keep them out the entrance is covered with a 'mouse guard' – a zinc strip perforated with $^3/_8''$ (9.5 mm) round holes which a mouse cannot force its way through.

Brood box

In simple terms a hive is composed of a floor, boxes for the bees to live in, and a roof on top. There are two types

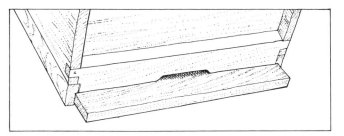

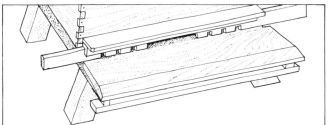

The entrance block of the National (above) can be turned to give a small opening, close the opening, or be removed completely. In winter it must be removed and replaced with a zinc mouse guard as shown.

The entrance blocks on a modern WBC (above right) are similarly adjustable. The small, square cut-outs in the blocks should suffice to keep out the mice in winter, though a mouse guard (right) is extra safe.

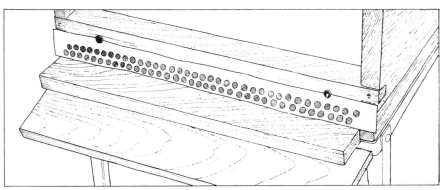

of boxes – brood boxes and supers. They work on the same principles and are constructed in the same manner; the difference is that brood boxes are deeper and have a different role.

Each hive has a single brood box, though that is complicated by the term 'brood and a half'. The brood box is the bees home, accommodating the frames on which the queen lays her eggs. The depth of a brood box is the same on a National or WBC (8⅞″in) [22.5cm], but while the former will take 11 frames the latter is one frame shorter and will only take 10 which is why they are not compatible. Both are 'bottom space' boxes, meaning that there is ⅜in (9.5mm) space allowing bees to run to and fro beneath the frames and up the sides; on a 'top bee space' box they would run across the top of and down them.

Supers

'Super' is Latin for *on top of*, and the supers are an exact fit on top of your brood box. National and WBC supers are 5⅞″in (15cm) deep, taking 11 and 10 frames respectively.

A super takes the frames which act as the bees' food and honey store. In

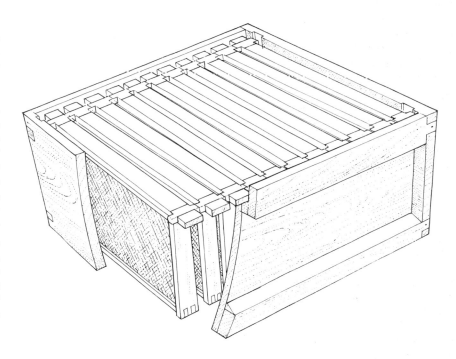

Cutaway view of a National 11 frame brood box. Brood boxes of WBC, Langstroth, Dadant, etc are similar, the main differences being in dimensions. The frames fit on runners with sufficient 'bee space' beneath.

the first year of beekeeping you may well only need one super complete with frames to go on top of your brood box. If you have a National you will also need an empty super to help with feeding the hive, storing your notes, etc. By the second year you are likely to need a total of at least three supers, plus enough frames for all of them – No. 1 for your brood and a half, No. 2 for your honey, and No. 3 to change over with No. 2 (and vice versa) when you're spinning off the honey.

However you may not have the time to take off the honey super every time it's full. In that case you will need more supers so you can leave two or three piled high on the hive, giving you more time before the bees fill them with honey. We are talking about a good year!

Lifts

The lifts are the outer walls of a WBC. Underneath it is really much the same as a National; outside it is a prettier version which probably has better insulation properties, though the bees don't seem to mind either way.

In your first year with a WBC you are likely to need three lifts – one to cover the brood box; one to cover the super; and one to give the same area of open space in the top of a hive, as with the empty super on top of a National.

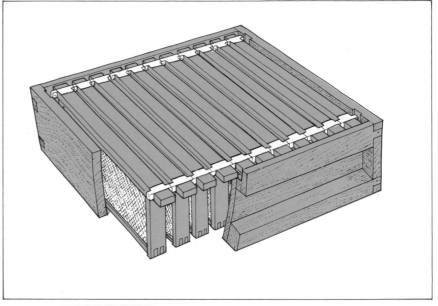

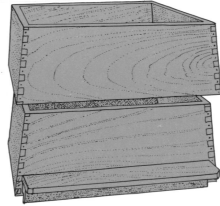

Above: A National super is the same dimensions as the brood box but is shallower with shallower frames. If buying home-made equipment check that brood box and supers fit together with no gaps for the bees to escape. Left: The lifts of a WBC double wall hive give the bees extra insulation and extra space outside the boxes. Lifting them off and on makes working a WBC more time consuming than a National; nor does a WBC transport so well.

BROOD AND A HALF

If a hive has a brood and a half it means it uses a brood box and a super to house the brood which is the egg laying area of the queen. It is called a brood and a half because the super has half the depth of the brood box and therefore has half the area of frames.

One of the skills of the beekeeper is to control the laying of the queen to maximise the use of each frame. As winter sets in a colony will have died back to around 10,000 bees, clustering to four or five frames. It is warmer and safer in the enclosed space of the single brood box.

However at the height of the season both colony and queen may run out of

space in this single brood box. The queen can lay up to 1,500 eggs a day during May. Taking into account hatching times and the number of cells on a brood box frame, this means that she would need a minimum of eight frames, laying from corner to corner.

The unfortunate natural tendency of the queen is to lay in an arc and miss out the corners. The beekeeper can encourage her to make that arc cover as much of the frames as possible, but the areas she cannot cover will effectively mean that she needs the services of another five or six brood frames. Five plus eight equals 13. Ten full frames plus 10 half frames on a WBC (11 on

a National) equals 15, which gives the colony two spare.

The super for the extra half brood is put on when the brood box looks dangerously overcrowded with bees; something which usually will not happen in the first season since the bees are far too busy drawing out wax on the frames. It is important not to be in too much of a hurry to put on the half. If you do it too soon, the Queen will shoot straight up into it before she has filled the bottom frames.

Frames for the brood box

Frames are four sided wooden supports for the wax comb which the bees live and work on. Wax foundation is commercially made from pure beeswax to fit a brood box or super. It is available with worker or larger, drone size cells.

The foundation is generally wired to give it some structural support. This is vital to take the weight of the brood and stores in the brood box; and unless you want 'cut comb' – cut out and eat everything including wax, old legs, etc – is recommended in the supers.

It is the natural instinct of the bees to build out the wax foundation until the cells are deep enough for the queen to lay, and for them to store honey and provisions. This is called 'drawing out'. The wax making glands of a young worker start to work after two and half weeks, and those workers are then assigned to draw out the foundation to a depth roughly in line with the width of the top bar of the frame, leaving an approximate ⅜in (9.5mm) passage for the bees to pass between each frame.

The frames are held the correct ⅜in (9.5mm) distance apart by removable metal or plastic ends. You can buy frames with specially shaped 'self-spacing' side bars, but it is more difficult to get a good tight fit. Metal ends are easier to handle. They allow easy access to each frame; or you can pack them in tightly (useful for transport) by opening out the flaps on each end frame.

It is important to keep track of the age of your frames, and for this reason frames in the brood box are coded with coloured ends, which we would recommend doing from the second year on. The colours are international and run in a five year colour system, starting with white:

Years ending in 1 or 6 – White
Years ending in 2 or 7 – Yellow
Years ending in 3 or 8 – Red
Years ending in 4 or 9 – Green
Years ending in 5 or 0 – Blue

The queen should also be marked with special paint, using the same colour coding to show her age.

You can buy plastic ends in the right colour, or failing that use coloured drawing pins. Only use the colour on

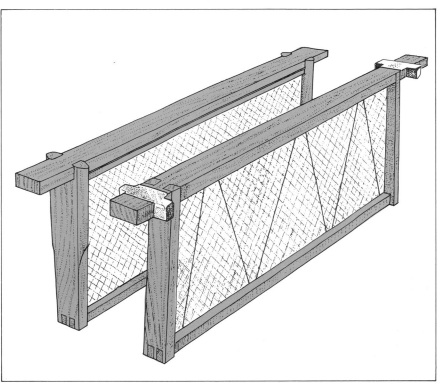

Above: Removable metal ends give the correct spacing for the frames. Self-spacing frames (left) are an alternative with shaped side bars.

Below: Drone foundation (top) has larger cells than worker (bottom). It can be used in honey supers, but this is not recommended for the novice.

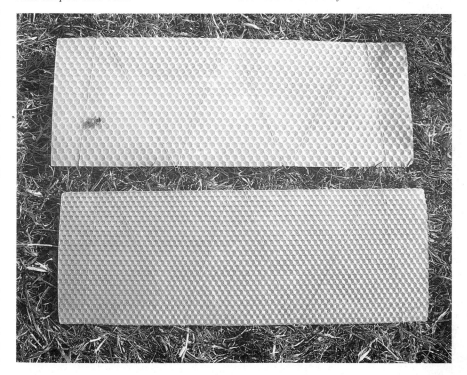

one side of the frame, with the regular silver metal ends on the other. This means you always put the full depth brood box frames back in the right way round so that the drawn out wax fits correctly together.

In the supers or brood and a half super this does not matter. The honey supers are always being spun off, and the brood super is likely to be moved on and off the hive as required during the season.

Frames for supers

Apart from being shallower (5½in) (14cm) frames for supers are basically exactly the same as for the brood box.

To start we would recommend that you only buy wired foundation which will also allow you to use these frames in the brood box super. Frames for supers should always be fitted with narrow, silver coloured metal ends. You can get wide metal ends, but these are for more advanced beekeeping.

Don't make the mistake of buying drone rather than worker foundation at this stage. The former is recognisable by having much bigger cells.

Glass quilt/Crown board/Clearing board

More confusion with three versions of a very similar item. You need a cover for the top of the brood box or super beneath the roof of the hive, and for most of the season we would strongly recommend you use a glass quilt. At other times you will need a crown board/clearing board which is effectively one and the same thing.

The glass quilt is a glass cover set in a wooden frame, with a single opening for feeding the bees in the central bar which can be closed with a Porter bee escape. The glass quilt was invented some years ago by Arthur Chitty of West Sussex, and it keeps the bees in and lets you look at them with impunity.

A crown board is the glass quilt's predecessor – the same thing in wood, but with two openings for bee escapes. Its disadvantage is that you can't watch the bees through it; its advantage is that it absorbs moisture and doesn't cause

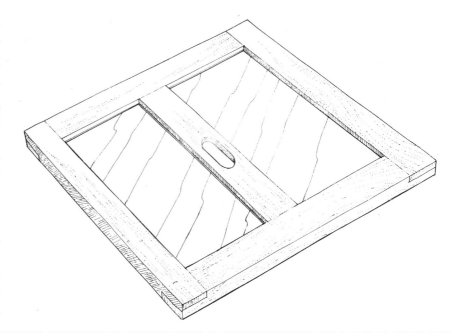

Above: The glass quilt allows you to lift the roof and watch your bees. The feeding hole can be closed.

Below: The underside of a crown board with a single bee escape in position. Most have two feed holes.

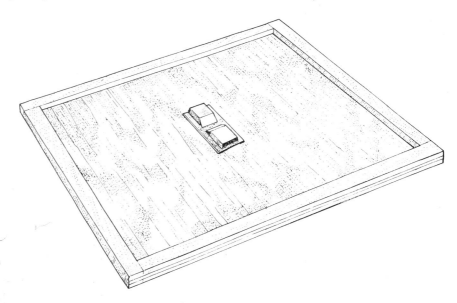

condensation as the glass quilt does in late autumn or winter. To overcome this condensation some beekeepers insulate their glass quilts with polystyrene; the trouble is that the bees tend to eat it, so we would recommend changing to a crown board for the colder weather when there is not much to see anyway.

A clearing board can be a glass quilt or a crown board, whichever is spare or handy. It is used to clear the bees down from the honey super which you wish to remove, and this is achieved by

Right: National hives and nuc boxes at an apiary. All the components are interchangeable. Note the hive stands.

means of blocking the opening (or openings) with a Porter bee escape. This ingenious device is fitted with a spring which leaves enough space for two workers at a time to get down into the next box, but they can't get back up.

All the workers will want to go back down to the queen once they have deposited their supplies in the honey super, and after 24 hours of this downward migration the top super should be empty of bees, allowing you to steal it away with relative impunity.

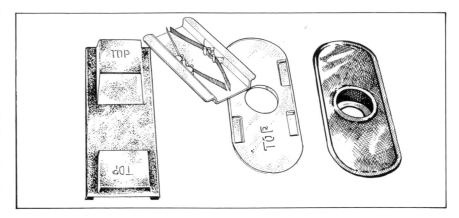

Queen excluder

You will only need a queen excluder in your first year if you get past a brood and a half. With the time taken to draw out all that new wax foundation the bees are unlikely to get that far, but you may be lucky.

The queen excluder is like a cattle grid which keeps the queen and drones down while the smaller workers can pass between the bars. Thus they can draw out and fill your honey super, but she can't get up there and start laying in it.

The old zinc kind of queen excluder is not recommended. The modern metal rod variety is much kinder to their wings, but they still don't like the cold feel of it and are unlikely to move up into the new super if it only contains undrawn frames. To lure them up you have to put in some drawn-out frames, but more of that in pages to come.

The roof

The roof of the hive keeps the rain off, and leaves some necessary space for ventilation over the top box of the hive.

Most roofs are made of wood, covered by thin alloy for extra protection against the elements. The problem is that water clings to the alloy, and bees that settle on the wet roof can drown on it. To overcome this you should roughen the surface of the alloy – use a scattering of sand on a wet coat of varnish, or you can buy proprietory non-slip deck finishes designed for yachts.

The National roof is available in two

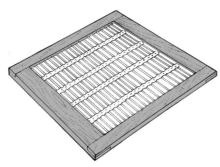

Above: Different types of bee escapes which fit in the feeding holes of a glass quilt or clearing board. They are designed so the workers can go down through them, but can't get back up. The Porter bee escape (centre) is a very popular design made in plastic.

Left: The old kind of queen excluder (above) is a flat sheet of zinc with holes punched through. This tends to distort, and can rip the wings of the bees. The modern version (below) with metal rods is much better, though the bees dislike its cold feel.

Below: The WBC (right) has a characteristic cottage style roof and outer lifts, but is otherwise similar to the National (left). Roofs should be given a rough finish to give bees some grip on the surface when it's wet – on a slippery flat roof they can drown.

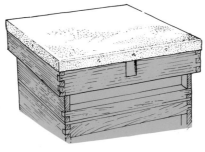
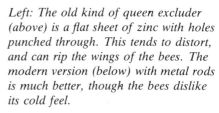

depths. The 6in (15cm) roof is less likely to blow off in exposed situations, but is somewhat clumsy to handle; the 4in (10cm) version is fine for town or suburban use.

Finally, it you want your National to look more like a WBC you can get a WBC style roof to fit it. However one of the advantages of a National or any of the more box shaped beehives is that it is not so obviously a beehive, and is less likely to worry neighbours!

The bees

Amazing though it may seem, there will be 50,000 or more bees in a healthy beehive at the height of the summer. In winter the number is trimmed down to a more manageable 10,000 or so to see them through the cold, unrewarding (for them) months.

50,000 bees may seem to equal 50,000 potential stings, but so long as you treat them properly they have no interest in harming you. So unless you are dressed up in protective clothing never go close to the front of the hive where you will block their flight path – if a bee leaving the hive flies slap bang into you she (all the flying worker bees are female) will understandably be put out.

If a bee alights on you don't worry. Whatever you do never flap your arms around and show signs of panic. She will assume you are a deranged beast and promptly sting – the last thing a bee wants to do since her sting is held so firmly by rough human skin that it takes half her abdomen with it and she effectively commits suicide. She does not have this problem with softer skinned animals and insects which are her normal enemy.

If bees seem to be taking an unhealthy interest in you walk away sedately and shown no signs of fear. If by any chance you are stung it is not that unpleasant, but remember that the sting leaves a scent which will guide other bees to the 'enemy'. The answer is to make an orderly retreat; or if you are experienced give the sting a puff with your smoker which will put off any of her friends and let you continue what you were doing.

TYPES OF BEES

There are three castes of bees in a hive – workers, drones, and the queen – all of whom are interdependent. The queen lays the eggs, the drones are the males who ensure the continuation of the species, and the workers feed the hive. They all have very different roles and characteristics.

Her Majesty

There is one queen in a colony and she will usually remain three years. After that the workers usually create a new queen, and she will end her time by being killed off or swarming to make way for her successor.

Calling her 'queen' is something of a misonomer, for she is basically no more than an egg laying machine. She starts life by being reared in a special queen cell, emanating from an egg laid by the then reigning queen who is to be replaced. Heavy feeding with royal jelly makes her grow very rapidly, and she emerges fully grown after 16 days.

There are likely to be a number of queens growing in this way. Some will be stung to death before emerging to reduce the numbers; survivors will leave with swarms ('casts'), set up new homes, and then be mated with male drone bees.

This brief fling with the drones will last a queen for the rest of her life during which she will be expected to lay something around half a million eggs! She commences this task within a couple of days of mating, and from there on does nothing but lay, lay, lay in the spring and summer high season, gradually running down in the autumn, stopping totally by December and starting again in a small way in January depending on the weather.

Worker Bees

There is a single queen in a colony and a few hundred male drones, and the rest of the numbers are made up female workers. 'Workers' sums up their lives, for there in no respite in working in the colony and their life expectancy is proportional to the amount of work that's going on – six weeks from birth to death in the busiest days of summer, and up to six months at other times of the year.

The tasks given to the workers are many and varied, and generally are related to their age or 'seniority',

though they probably don't see it that way. For instance the duties as they progress through life could be cell cleaning and incubation for the 0–4 days olds; then feeding larvae; making their first test flights (6–10 days); wax making and comb building; receiving and storing nectar; entrance guarding and debris clearing (14–21 days); and thereafter foraging for honey and pollen and collecting propolis.

It is the foraging trips that really take it out of them, and if they are kept in the hive by bad weather their life expectancy is lengthened. In the autumn and winter they carry their own food reserves, using them to make brood food when the queen starts laying once again.

Drones

Drones are recognisable by being larger than workers, and have the advantage (for us) that they have no stings. They are born from unfertilised eggs laid by the queen in special outsize drone cells in the spring, and the appearance of these unusual cells is insurance against the impending death of the current queen and can be an indication that the colony is preparing to swarm

Drones start to fly about a week after coming out of their cells and after a further week are virile enough to mate with a queen, which is after all what they are there for – these boys don't do much work, but they do ensure the continuation of the species!

During the spring and high summer drones are welcome visitors to all hives, but come late summer they are considered surplus to requirements since their mating prowess is not required and they are just a drain on the colony. They are driven out of the front door of the hive, stung with one wing bitten off, never to fly again – the end of their comparatively short three or four month lives. In this case the men are definitely not in charge!

Honey Bees in Detail

The correct name for the honey bee is *'Apis Mellifera'*. The UK honey bee used to be the Black Bee which was almost wiped out by the 'Isle of Wight disease' (probably Nosema) just after the First World War.

After that the British had to turn to imported bees, and in particular Italian bees became very popular. They were prolific and quiet, but their one disadvantage was that colonies which produced so many bees meant less honey for the beekeeper.

Today's British bees are largely descended from Italian and other imported bees where the tight import restrictions allow. Most recently 'golden' New Zealand bees have become popular. However even though the genes of bees may be controlled in an artificial environment or by artificial insemination, there is no way that you can prevent bees becoming mongrelised when they swarm. All you hope is that they retain their good habits, and lose their bad ones.

Bad habits include bees which 'follow' – ie. bees which follow people and perhaps sting them. If a hive has bad habits it can be interbred with a hive with good habits in the hope of only retaining the latter; in the case of the Africanised 'killer bees' (see page 168) this has proved impossible with disastrous results.

Good habits include bees which are reasonably docile, stay still on a frame when you move it, don't 'follow', winter well, are good honey gatherers and reasonably prolific – enough to build up a good size colony, but not so many that they consume all the stores.

Bee diseases

Import restrictions are to prevent the introduction of disease, most notably the Varroa mite which has spread through most of Europe depleting colonies on its way. It has reached France, and is no doubt only waiting for the Channel Tunnel as its passage to the UK! On the other side of the Atlantic it has recently been identified in the USA, and American imports of queens to the UK are now banned.

We are already afflicted by American and European Foul Brood. These are highly contagious bee diseases which spread rapidly. American Foul Brood is the most dangerous, and any afflicted colonies must be immediately destroyed by burning the bees on their frames which harbour the disease. In serious cases of European Foul Brood the same

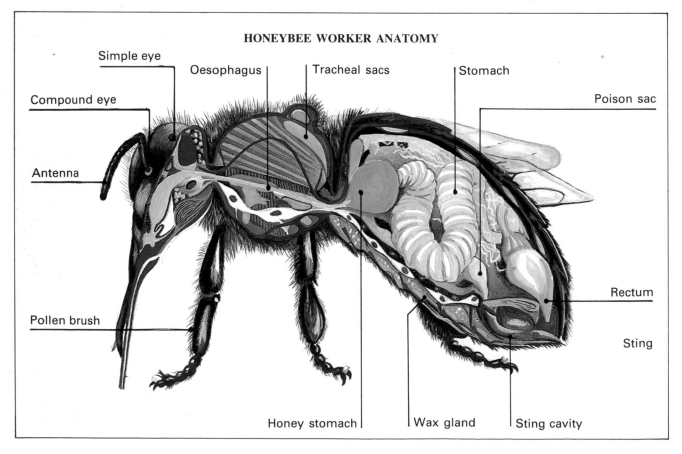

HONEYBEE WORKER ANATOMY

Simple eye
Oesophagus
Tracheal sacs
Stomach
Compound eye
Poison sac
Antenna
Pollen brush
Rectum
Sting
Honey stomach
Wax gland
Sting cavity

QUEEN

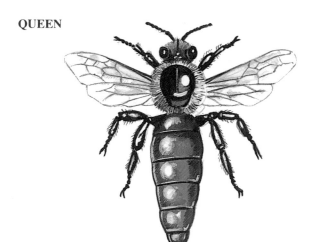

DRONE

WORKER

Identifying the bees in your hive. The drones (above) are easy to tell from the workers (left), but the solitary queen (top left) can be very tricky to find. Being able to spot her is highly important when you start to use anti-swarming techniques such as requeening, and for this the queen is marked with a blob of colour appropriate for her year – red, white, yellow, blue, green.

Drones, workers and queen are all interdependent, and it is usually the poor male drone which is most overlooked and underrated. Each fine summer afternoon, drones from a mixture of colonies meet and fly around for a couple of hours, hoping to meet a new queen on her mating flight. Great numbers of drones will chase her, and about seven will overtake and mate with her in turn while flying, giving her all the sperm she will need for her life. Each of these drones falls to the ground paralysed. . . .

drastic measures are necessary.

The surest way to ward off Foul Brood is to keep your hives clean and in good order. In the event of an outbreak you must report immediately to the Ministry of Agriculture.

Diseases and how to identify and control them are dealt with in more detail on page 170.

Not honey bees!
There is sometimes confusion about how a humble bumble bee relates to a honey bee. The answer is not very closely.

The bumble bee is a marvellous polli-nating machine with a longer tongue than a honey bee. It lives in small colonies of some 200 bees, preferably in rubbish tip, and only the queen survives from year to year. The bumble bee can

sting, but it's unusual, and the bumble bee colony is only interested in producing honey for its own use with none left over for humans.

Lastly there is the tiresome wasp which is no relation at all. It has a different sting, lives in a nest like a papier mache football, and is inclined to rob honey from bees if it can. For us humans it is just a nuisance.

A mass of workers busily covering a frame. At first this sight may be intimidating, but you soon realise that they are not in the least bit interested in you so long as you don't threaten them. Careful and confident handling comes quickly with a little experience; many long term experts prefer to wear no gloves.

Protective Clothing

You sometimes see pictures of beekeepers holding up frames of bees without a hat, veil, gloves or any protective clothing. Some experts really don't need protective clothing and others just like to be a little macho, but rest assured that they are either real experts or they are foolish.

You don't need to dress up to take the roof off and peer down through the glass quilt, but do tread quietly and carefully and be sure not to antagonise the bees. However if you are going to lift the glass quilt or do anything that exposes the frames we would always recommend dressing protectively. The bees don't want to sting you, but your own ineptitude might encourage them to do so. This is particulary true when they are guarding their honey at the end of the honey flow in late August, particularly in their second year.

The main requirements of protective clothing are that you must be properly covered, and you must be self-confident and comfortable in what you are wearing.

Priorities

Number one priority is to protect your face and head. Due to the pollen collecting hooks on their legs bees often get hopelessly entangled if they land on a head of hair or in a beard, and when you panic and try to get rid of them they sense your fear, panic and sting you in return. Your face contains some highly sensitive areas which don't take kindly to stinging – a sting on the mouth or eyelid can result in severe and possibly dangerous swelling.

The simplest protection for this area is a round hat with a veil that hangs well away from your skin. Bees have a natural tendency to go up, so take care that they cannot get up inside the veil.

The best kind of bee suit is without a doubt the overall, complete with goatskin gloves and boots worn outside the trousers to stop bees running up.

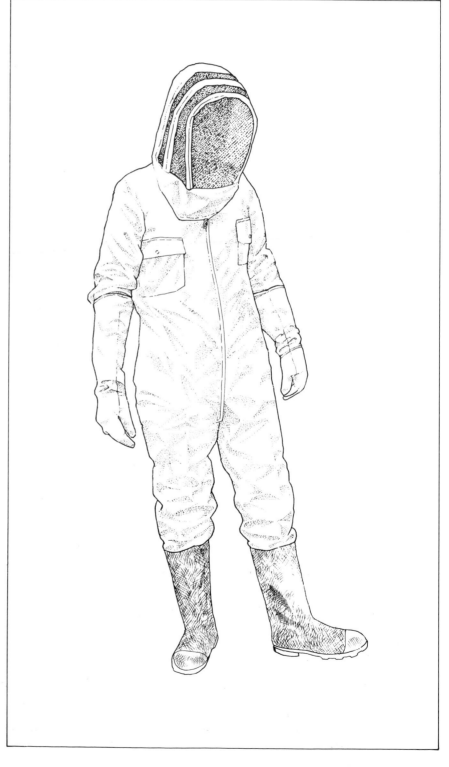

This can only happen if you don't adjust it properly, but if a bee does get inside:

1 Remain calm.
2 Move well away from the hive.
3 Take the hat off to remove the bee, or if necessary squash it between gloved fingers.

Never attempt to remove the hat while standing by the hive. The bee may sting you as you do so, immediately attracting other bees to sting you in the same area.

The next priority must be gloves. Washing-up gloves are sting proof, but they may expose the wrists and are sweaty. We would recommend buying good quality goatskin beekeeping gloves with gauntlets which offer complete protection.

With the hat, veil and gloves you can make do with ordinary clothing, so long as you remember to cover all exposed areas and take note of the fact that bees go up. A bee flying up a loose trouser leg may appear amusing to others, but if it's your trouser leg you will wish that you had tucked it into boots or thick socks, or some make do with bicycle clips.

Remember also that the bee sting will penetrate tight jeans, thin socks, etc. Loose fitting, relatively thick or tough clothing is necessary to keep that little barb away from your skin.

The well dressed beekeeper
The best kind of beekeeping suit is an all-in-one overall incorporating a hood which can be unzipped and flipped back off your head. Made in polycotton we would definitely recommend this type for peace of mind, comfort, and a good selection of pockets for stowing hive tools and other small items. It is also of course the most expensive kind of beekeeping garment – a good choice for a second year beekeeper's Christmas.

Next best is an anorak incorporating a hood, but remember to tuck in the waist or pull it in 'bee-tight'. To complete the ensemble you will of course have goatskin gauntlet gloves, and matching white wellingtons have to be a 'must' for the upwardly mobile!

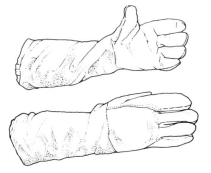

Above: Being chased by a swarm of bees! It doesn't happen like this, but occasionally quite a few might decide to 'come and get you'.

Left: Bee gauntlets make for safe handling of bees.

Below: Two kinds of veil. The type which can be attached (usually by zipper) to a bee suit is preferable.

Full beekeeping regalia is quite hot and stuffy to wear in very warm weather. Lightweight veils are available, but the British beekeeper seldom has much of a problem in that respect – on the warmest days of summer when putting on a clearing board or taking off a super you should only work on the hive in the cool of the evening when the bees are safely 'tucked up for the night'.

Great grandad and great grand-daughter meet with about 100 years between them. Modern clothing gives complete protection against bee stings, if worn correctly. Boots mut go outside trousers so the bees can't climb up; gauntlets should give a double covering at the wrists so bees can't sting through; and the hood must be secured so there is no chance of a bee finding its way in which is an alarming experience! We would recommend the complete one-piece overall as giving far and away the best protection; plus it's comfortable and easy to put on and take off.

There's also a marked difference in smokers in the pictures, though in this respect we haven't made much progress. The bellows smoker in general use today can be tricky to light with a tendency to stop burning at inopportune moments. Success with it mostly depends on getting suitable fuel, but a more modern and efficient design is long overdue!

Hive and Frame Construction

The favourite wood for the outside of a beehive is either cedar, or a good quality softwood such as Ponderosa pine or Scandinavian redwood. The inside of the hive which is not exposed to the weather can be made from any softwood. Thus a top quality WBC will have cedar feet, floor and lifts, with softwood boxes and frames inside. Since the boxes of a National are exposed, the best should be cedar with softwood frames.

A cedar hive should last a lifetime. You do not paint it for it is a natural wood which dissipates moisture, although treatment of the outside with a good quality preservative which is non toxic to bees is advisable every few years.

If your hive is made from one of the other soft woods you must paint it (non-toxic) or treat it with preservative every year to prevent the wood warping or rotting. In New Zealand they dip their softwood hives into hot wax to preserve them.

Making up a hive

You can buy a new hive ready made, or 'in-the-flat' which saves about 10%. If you are a handy woodworker you can of course make your own from scratch.

If you buy 'in-the-flat' the pieces all fit together like a Meccano kit, and there is a lot of reward in making up your own gear. The tools you will require are a hammer, a bradawl, a set square, waterproof wood glue, and galvanished or sheradised nails. A National is fast and very easy to make; a WBC takes longer due to the more complicated lifts, floor and roof – allow about a day.

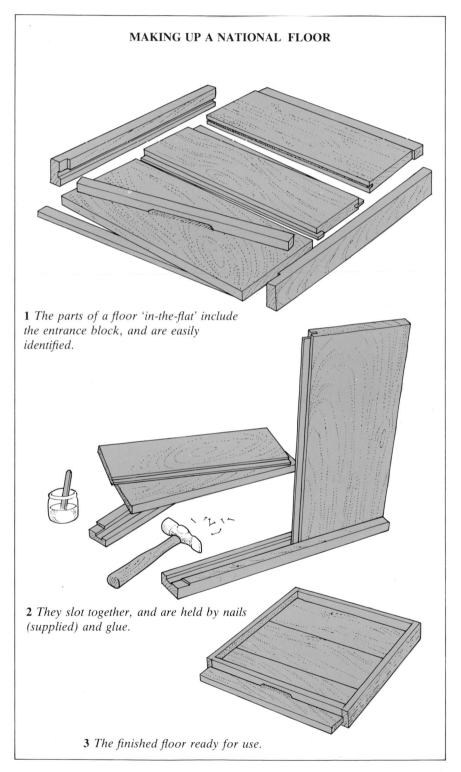

MAKING UP A NATIONAL FLOOR

1 The parts of a floor 'in-the-flat' include the entrance block, and are easily identified.

2 They slot together, and are held by nails (supplied) and glue.

3 The finished floor ready for use.

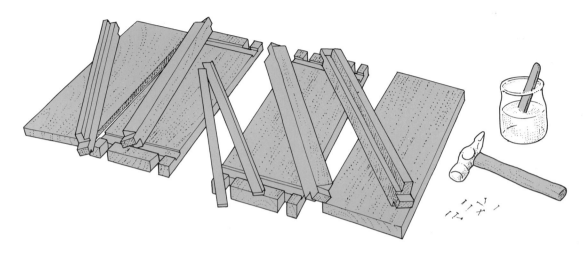

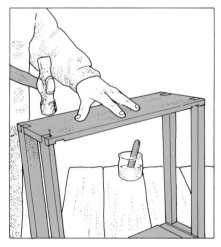

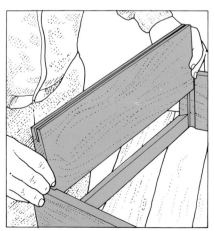

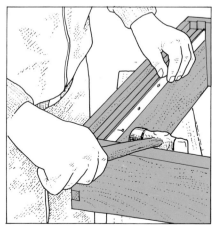

You can buy the main parts of a hive in pre-cut 'flat' form ready to be assembled. Nails are supplied; all you need is a mallet, hammer, and wood glue. It is interesting to make your own equipment in this manner, and it will save you a little money which is always worthwhile!

Assembling brood boxes and supers is exactly the same procedure.

1 *Use the mallet to join the two complete sides of the box to two bars on both sides as shown. Use the pre-shaped cutouts; make sure the sides are all the right way up; and glue the joints before hammering home.*

2 *Drive nails into place to hold the joints in position.*

3 *Slide in the two inside end panels which will hold the runners for the frames. Use glue to hold them; they will keep the box in a rigid square shape.*

4 *Nail the two runners into position.*

MAKING A NATIONAL ROOF

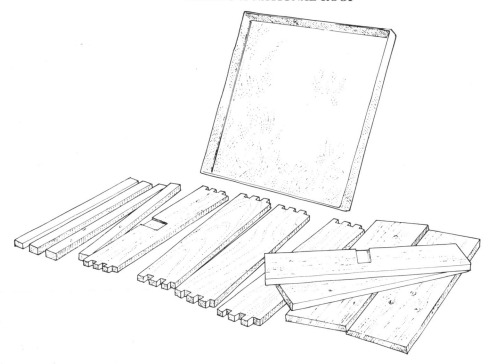

Making A National Roof: The four sides of a roof supplied 'in-the-flat' are first fitted together using the cut-outs, held in position by wood glue and nails.

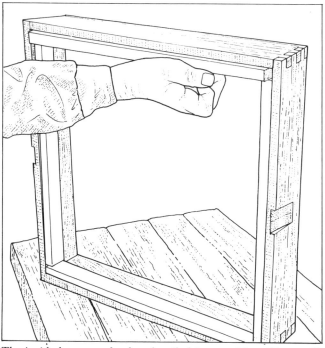

The inside bars are glued and nailed into position. These will rest on the crown board, and help to reinforce the roof. All the parts are supplied in a single package.

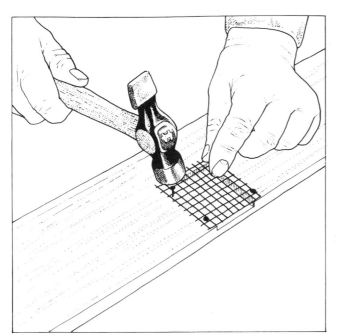

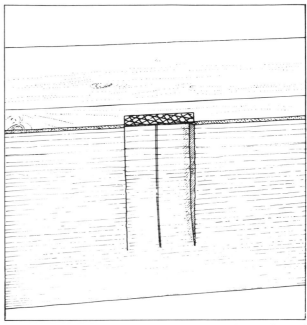

The two end roof slats have ventilation holes which are covered over with mesh to stop the bees escaping. Don't nail the mesh over the sides of the roof.

Looking sideways-on at one side of the roof with the end roof slat nailed down. The bees get plenty of air from the ventilation hole, with the mesh preventing their escape.

The roof slats are glued and nailed in position. All parts are pre-cut, and no shaping or cutting is necessary. Only basic woodworking skills are required, plus a few tools.

Finally the metal roof cover is dropped over the top and nailed through to the wood. It will need no upkeep, but should be given a rough finish using sand and varnish.

Making frames

As the bees replace themselves, so the frames are the one part of a beekeeper's equipment which need regular replacement on a five year rotational basis.

You can buy them for around 50p each made-up, but it's more fun to make your own – it's also preferable since you can store the parts and make them up at the beginning of the season as required.

The tools you need for frame making are a small hammer or 'Brad Driver' and gimp pins. The frames are made from softwood with wax foundation, and if a frame needs mending any of the parts can be simply replaced as can the wax foundation.

MAKING UP FRAMES

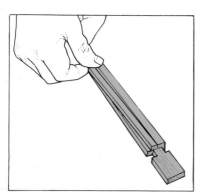

Start by prying the loose wood strip from the top bar.

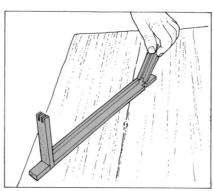

Use a mallet to tap the side bars home into the ends of the top bar.

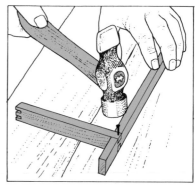

Hammer in gimp pins to hold the side bars in place on the top bar.

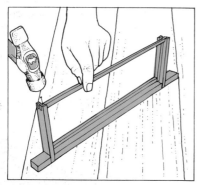

Carefully slot in the first bottom bar, and hammer in gimp pins as shown.

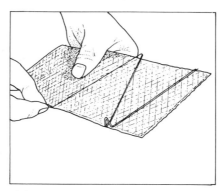

If the foundation is wired, bend up the three loops at 90 degrees.

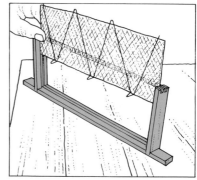

Very carefully slide in the wax foundation down onto the top bar.

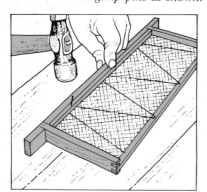

Insert the loose wood strip, hammer in gimp pins to diagonally to hold it.

Finally put in the second bottom bar, and secure it with gimp pins.

Metal ends must be added before using the frames.

Buying Secondhand

There is a lot of secondhand beekeeping equipment available. Beekeepers die, or can't handle it though ill health or other pressures – sometimes they suddenly become allergic to stings.

Secondhand Nationals and WBC's

A WBC is fine if you have one or two hives, but becomes a nuisance to handle if you have more. For this reason many beekeepers change to Nationals, and consequently there is always plenty of secondhand WBC equipment on sale. The simpler National is frequently home made, so is also widely available.
1 If it is a home made hive check the dimensions of the boxes. They may be a poor fit which lets out the bees, and also not interchangeable with correctly sized boxes and components.
2 Check that the wood is in good condition. On a National check for woodpecker holes.
3 Check that the inside of the hive has been kept reasonably clean, and if not leave well alone. However good it appears you must sterilise all secondhand equipment with a blow torch and a scraper (use the sharp end of the yellow tool). Get into the corners to get rid of all moth eggs and possible disease.
4 Check whether the hive is top or bottom space. Most are bottom space, but you can't mix and match in a hive.

Secondhand Frames

If there is disease in a hive it will most likely be harboured in the wax or wood of the frames. Why take the risk of buying secondhand? Frames are the cheapest item of new beekeeping equipment, and brood frames should be thrown out every five years on a strict rotational system. Old super frames can be sterilised with Acetic acid, but this corrodes the gimp pins so that they eventually fall apart.

Our advice is: DON'T USE SECONDHAND FRAMES – BURN THEM.

However if you are offered second-hand unused frames in the flat, that can be a very good buy. Check that you have the metal ends; check that the frames are the right size for your hives and the wax foundation is the right size for the frames, as well as being worker or drone as applicable; and store the foundation in dry plastic bags.

Other secondhand equipment

Clothing: Guidelines are as for new. Make sure there are no holes, particularly in the netting. Check gloves carefully.
Smokers: Check bellows. Remember that the tin plate corrodes.
Hive tools: Check they are not bent or blunted.
Nuc or travelling boxes: These are comparatively expensive new (around £30 and £18 respectively) so can be very good buys secondhand. They hold 4 – 6 frames accommodating up to 10,000 bees, and are particularly useful for temporarily accommodating swarms. Sterilise as for a hive.
Honey containers: The lining comes away from galvanized honey buckets so they are not a particularly good second-hand buy. Plastic buckets and feeders should last indefinitely, though check that mesh feeder filters are stainless steel, not tin plate. Glass jars are well worth getting hold of as a cheap secondhand job lot.
Honey extractor: May well be included if you buy a complete secondhand package of equipment. The older tin plate versions are fine for domestic use, but under EEC regulations an extractor must be plastic or steel if you wish to sell the honey.

You should see the extractor working. Wash it thoroughly with cold water to remove any wax crumbs, and winter it by painting the inside with olive oil.

Other more exotic gear such as solar wax extractors can also be found secondhand, particularly if you buy a complete package.

Secondhand beekeeping equipment for sale! There are bargains if you know what to look for, but that requires knowledge of beekeeping and prices.

The First Year: Siting Your Bees

Having examined what you need to start beekeeping, you then must decide where you're going to put it all. For those with wide open spaces and plenty of storage the solutions tend to be easy, though hives should always be easily accessible; for those with more confined space the solutions are more tricky, but it is still possible to keep bees in a tiny garden – producing plenty of honey and keeping the neighbours happy!

Hives in the Stony Meadow apiary leave plenty of room for people and bees. One can walk in complete safety behind the hives; or pass in front, far enough away to be clear of the bees' flight path as they head off for their forage. The neighbours also live in harmony with the bees, with the occasional early summer swarm usually settling in the apiary. Anything wrong? Having hives in a line will inevitably lead to bees 'drifting' from hive to hive! For the hobbyist with more than one hive we recommend having entrances at different angles to avoid this.

Selecting a Site

When selecting a site for your beehive, you are looking for an area which is:

1 Relatively flat and dry.

2 Not exposed to the full force of the wind.

3 Clear of damp undergrowth.

4 With a fair amount of forage close to hand – eg. don't stick your bees in the middle of a cornfield.

5 With a flight path well clear of humans so that the bees are safe and not a nuisance – particularly important in town.

6 Easily accessible for carrying and storage – important in a big garden.

7 Undisturbed and quiet.

8 Not in the same field as disruptive elements such as horses which like to kick beehives with their hooves.

You need to decide where the hive or hives are going to go; how many if more than one; and the direction of their entrances. While bees can be accommodated in the country with ease, they can also be accommodated in the town just as well. A flat garage roof or roof garden will do well if it takes the weight, but with swarming you must know what you're doing to control the bees and prevent them being a nuisance and possible danger to neighbours.

The illustration opposite shows a hive in three potential positions in a small garden, and what is right and wrong with them:

Site A

● Bees will fly straight into anyone on the lawn.

● When going through the hive, no one could use that part of the garden near the house.

● Bees will stain all the washing bright yellow with their defecation.

● The hive entrance is well protected from the prevailing wind – probably the only good thing about this position.

Site B

● The prevailing wind blows into the entrance.

● The bees fly straight across the path.

● Fast moving, excitable children in the playground immediately adjacent will result in mass stings!

Site C

● Plenty of room for a second hive.

● Entrances away from the prevailing wind.

● The bees dance to decide where to go, and then fly up and away. You need some obstruction to take them up steeply if there are humans around, and since the bees take off in a circular spiral they need to be prevented from horizontal flight on all four sides in a small garden.

Bushes, fences, trees, shrubs, etc will all usually do, as will chain link or chicken wire which the bees perceive as a solid barrier.

The adjacent 10ft (3.05m) wall should take the bees up and away, but in this situation it's so close that it will intimidate them if their entrance faces it. Therefore our sensible beekeeper has put up a screen of runner beans facing the two potential entrances. It is 5–6ft (1.5–1.8m) away and about 6ft (1.8m) high – just right to take the bees up and away on their flight path without bumping into anyone. A latticed fence or Jerusalem artichokes would do just as well, and this is only needed for the busy summer period.

Two Hives?

Always think in terms of having enough space for two hives. We would generally recommend getting a second hive for your second or third year. One hive can swarm or die out for various reasons, but with two you are far less likely to lose all your bees.

The golden rule with two or more hives is THEY MUST NOT FACE THE SAME WAY. If they do, the bees will drift from one to another, with the risk of spreading disease. Tired bees coming in from the honey flow will pack the first hive, while those further down the line receive less and less honey.

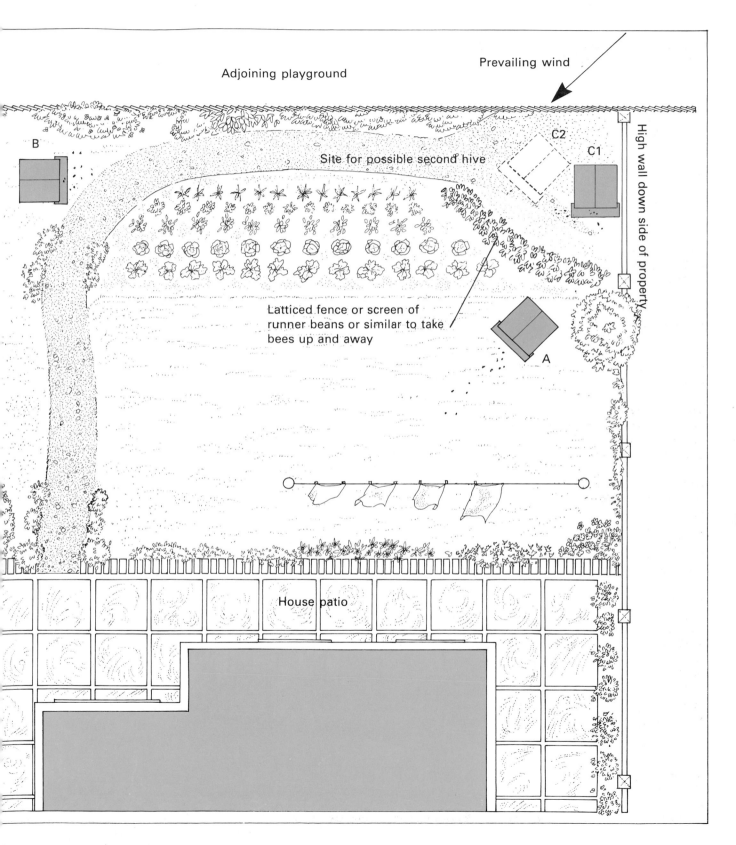

Adjoining playground

Prevailing wind

High wall down side of property

B

C2

C1

Site for possible second hive

Latticed fence or screen of runner beans or similar to take bees up and away

A

House patio

Installing the Hive

On this page we explain how to install a brand new hive in the position you have chosen, ready to accept the bees. The rules for an empty secondhand hive are the same, but be sure it is completely clean and disease free before introducing any bees. The old frames must of course be burnt, and new ones substituted in their place.

Moving a full hive of bees is another matter, and is dealt with on page 94.

Putting down the floor

First level the site, and clear away any weeds or undergrowth which will harbour damp. These should be kept clear of the hive all year round. If the hive is on a lawn, you can cut around it with the lawnmower, but wear a hat and veil when cutting in front of the hive. The bees don't mind the constant, steady din of a motor mower; they do not like sudden noises which they are not used to.

The best base of all is damp free concrete. Failing that we would always recommend standing the hive on a paving stone or similar blocks. If you have a WBC it's worth covering the bottom of the feet with roofing felt patches before putting the floor down. The National has no feet and its floor should never be placed straight on the ground. Place it on breeze blocks, or a metal or wooden stand.

Before positioning the floor, be sure you have considered where a second hive could be placed. Once the bees are settled in a hive, you are very restricted on how much you can move it due to the Rule of 3's – only move less than three feet or more than three miles. If you must move the hive a short distance (less than three feet), part of the entrance in the new position will have to correspond to where the entrance used to be. Furthermore if you are turning the hive to face a completely different direction (but never right round), you should only change it to that position in four small stages over

Migratory hives are often placed this way in fields, but we wouldn't recommend it for long term positioning.

The flying bees will tend to drift into the wrong hives, favouring the nearest to the forage.

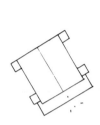 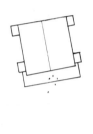 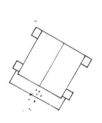

Placing the hives in more of a semi-circle is less confusing for the flying bees. The forage will get more equally

returned to the hives, and you cut down the chance of the spread of disease from one hive to the others.

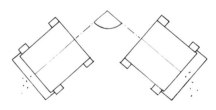

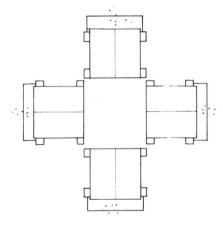

Above: Two hives are most easily worked placed as shown. The angle shown between them is about right for easy handling. Right: Four hives can be placed in a square allowing the beekeeper to work in the middle.

two days when the bees are flying. Otherwise they may not be able to relocate themselves.

Adding the rest of the hive

When positioning a new hive, the next stage is to put on the brood box.

The brood box should be filled with frames – 11 in a National and 10 in a WBC – with silver metal ends on both ends for your first year. Take care not to use the wide metal ends which hold

the frames further apart. They are for potential second year use, and are only for honey supers.

You must next decide whether to put your frames in the 'warm way' or 'cold way' – whichever way you must put them in the same way throughout the hive.

The warm way means having the frames parallel to the entrance. This makes it easier for the bees to repel attacks, helps keep the wind out, and is generally preferable for keeping a weak colony going through the winter. It does however slightly slow down a strong colony.

Unless your colony is weak, we would recommend putting in frames the cold way which is at right angles to the entrance. The wind blows in, but the bees can work faster in the hive since they find a far quicker route up to the honey supers.

Next put on your glass quilt. Before putting on the roof put on an empty super if you have a National, or an extra lift if you have a WBC. This gives you space to work over the bees, and is necessary when you need an enclosed space to feed them.

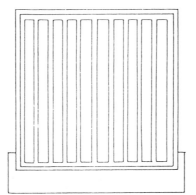 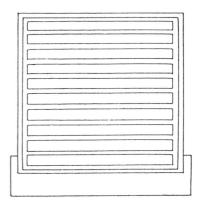

Frames are put in the cold (left) or warm (right) way. Ensure the frames in the supers run parallel throughout the hive or the bees will get muddled!

Below: Nucs of bees being prepared at a bee breeder's. You should take delivery of your nuc when it is ready to vigorously expand to a fully colony.

Installing the Bees

Here we are assuming you are getting a nucleus of bees in a nuc box or travelling box. As mentioned on the previous pages, installing a full hive is a second year subject covered under transporting bees on page 95.

When buying new bees from a dealer, you buy a 'nucleus'. These should be ordered in winter by January if possible, because supply of good, young bees is limited. You should expect delivery in late May or early June depending on the weather. A good nucleus (a young family) will consist of around five frames covered in some 10,000 bees, all raring to go. At least two of those frames should be solid with brood (sealed and unsealed) and the rest with the bees' stores made up of pollen and honey.

The bees will be collected (or can be delivered) in a nuc or travelling box – the latter is quite safe sent by train. You should expect to pay around £10 for each of the frames, with a returnable deposit on the box.

D Day

Your supplier will decide that the nucleus is ready when the weather has turned summery and the queen has started to lay flat out. He will contact you and will normally fit in with a convenient date. However the queen will be laying so fast that the nucleus could become uncomfortably over-crowded within a week, so if you can't take the bees before that time your supplier will sell them to someone else on his list.

If you are collecting the bees, arrange collection late in the evening or early in the morning. Otherwise half of them will be out flying at the apiary, and you won't get them. Alternatively if the bees are being delivered they can travel for up to two days in an unattended travelling box, but mishandling and overheating is a danger. As a beginner you want your first bees to arrive in good condition, and in a good mood!

Above: A travelling box is a handy beekeeping accessory for transporting bees. It can double as a nuc box.

Below: a nuc box will become useful in the second or third year for making up a small colony or housing a swarm.

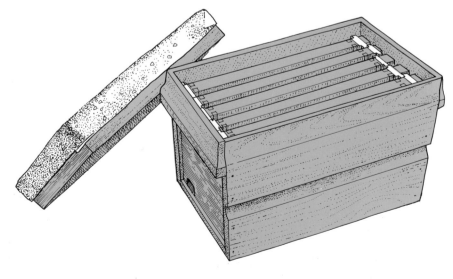

The neighbours

Whether you tell the neighbours is a personal decision, but if they're reasonable people it makes sense to get them involved from the start. If they are close neighbours you will have to tell them for safety's sake.

This also applies to your kids and their friends. Start them off so that they are confident and enjoy and respect the bees. It is well worth getting them suits so that they can join in – or one suit so that they can take it in turns is often even better.

Animals will learn to leave the bees alone. Once your dog gets stung it will keep clear. If you have a pony keep it away, since it will inevitably kick the hive. If you have been riding that pony, keep yourself away since bees hate the smell of horse sweat. They think it's that pony come to kick over their hive again, and will come out and sting you instead.

What you need

At this stage you should have ready:

1 Full protective clothing.
2 Smoker, fuel and a lighter.
3 A hive tool – the frames of the nucleus will be well glued together and a red one is best for this task.
4 A white or off-white sheet. If you drop the queen or any other bees you will see them and not trample on them. This is imperative for beginners.
5 A manipulation cloth and a soft cloth to partially cover the frames if necessary.
6 A bee brush or a long feather. We would also recommend you have something like a wheelbarrow to carry all the odds and ends in.

When you start

Take the nuc or travelling box down to the beehive. From now on we shall call it a nuc box, but the principles are exactly the same for the very similar travelling box.

Remove the roof, brood box, etc and place the nuc box on the floor, entrance aligned to entrance. There is no need to dress up for this operation since the bees are closed in the nuc box and cannot get out. Leave them that way so they can settle down. If there is a chance of rain, cover the ventilation holes in the top of the box with a sheet of plastic.

After around 30 minutes (time for a cup of tea) get togged up in full gear so that you feel confident. You are now going to let the bees find their way out of the nuc box, but they must have something to check their flight so they don't rush out and get lost – they are after all a bit hot and bothered.

A small bushy branch placed in front of the nuc entrance is ideal. Don't confuse them with a solid object or first time out they may zoom straight up over it and away. You can then remove the entrance block and the bits of foam rubber that are sometimes used to completely seal in the bees.

If you had rushed through to this stage the bees would all rush out in an agitated state; if you have taken it easy, the bees will be reasonably relaxed and will come out gradually.

Above: Place the nuc box on the floor of the hive, entrance aligned to entrance. Leave the bees for 30 minutes or so to settle down.

Below: Before letting them out, give them some kind of barrier so that they have to stop and 'think' before shooting off into the blue yonder.

Hiving the Bees

Leave the bees to their own devices until mid-day the next day which is the time when half the bees will be out flying on a fine day. You can then hive them if the weather is suitable – mild enough for you to go outside in shirt sleeves.

Unsuitable weather is:
- Rain
- Strong wind
- A sudden cold spell

Any of these three and you must leave the bees. They can look after themselves without being hived for up to a week.

To hive the bees, plan what you are going to do, and aim to do nothing else:

1 Get fully dressed and have all the necessary gear ready with your smoker lit. Go to the hive area.

2 Gently smoke the entrance of the nuc, then leave for three minutes. That's time for the bees inside to shout 'Fire!', dive down and start filling up with honey.

3 Lay your sheet out to one side a few feet from the hive site, and place all the gear on it including the various parts of the hive.

4 Pick up the nuc from behind. Carry it over to the sheet and place it down (B) with the entrance facing in a different direction so the flying bees can't find their way back in. Carry the hive floor over to the sheet and place the brood box on it. (If you put bees into a brood box without a floor they may fall out of the bottom!)

5 Place a cardboard box or skep on the hive site (A) to act as a temporary home for the flying bees. Like little robots they will come back and move into it,

in an apathetic state because they have lost the queen – 'their mum'. The skep should be upside down with one side propped up to make an entrance; the bees will then go up into it.

6 Take the lid off the nuc box and cover it over with the soft cloth.

7 Move the brood box and floor close by, with the entrance closed and all but a few of the empty frames removed. This gives you enough room to make a big space in the middle so you can lift the bee covered frames out of the nuc box one at a time, and slide them into the brood box without squashing the bees. The frames will be stuck down to the nuc so you will need to use the curved end of your red hive tool to free each end. Lift them slightly, and then grasp the ends between forefinger and thumb.

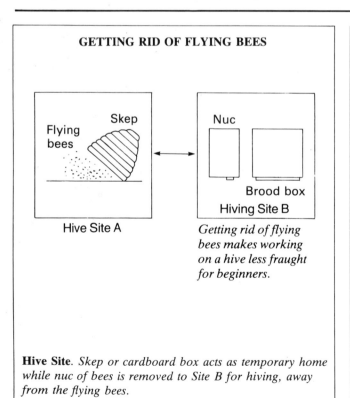

GETTING RID OF FLYING BEES

Flying bees — Skep — Hive Site A

Nuc — Brood box — Hiving Site B

Getting rid of flying bees makes working on a hive less fraught for beginners.

Hive Site. *Skep or cardboard box acts as temporary home while nuc of bees is removed to Site B for hiving, away from the flying bees.*

Hiving Site. *The nuc of bees is transferred into the brood box away from the hive site.*

Smoke the nuc entrance; wait three minutes before placing it on the sheet next to floor and brood box.

When you transfer the frames work quickly and methodically, lifting one frame at a time and leaving the rest covered by the soft cloth. This isn't the time to examine your bees for they are likely to be disturbed and unhappy by yet another change to their routine.

It is very important that the frames go into the brood box the way they came out of the nuc. They must be in the same order and face together the same way, for the bumps and indentations of the drawn-out wax have been designed by the bees to fit perfectly to maintain the brood temperature.

When all your bee covered frames are in the brood box, put in empty frames on either side to make up the total number – 11 in a National and 10 in a WBC. They should all have silver metal ends, and the ones on the end frames should be bent out to hold all the frames tightly in the box. Then cover quickly with the glass quilt which should have the Porter bee escape in place so bees can't come out of the top.

Putting on the glass quilt without squashing a few bees is tricky. The trick is to lay your soft cloth over the top of the frames (tea towel size) which forces the bees down since they don't like the weight on their backs. Place the quilt on top and whip out the cloth from under it.

8 Move aside the skep or box from the hive site (A). You may imagine the bees in it will be confused, but they will know where to go back to.

Carry the brood box and floor over to the exact hive site, facing the way it was originally. Then reassemble the hive, adding lifts if it is a WBC, an empty super if it is a National, and putting the roof on top after removing the Porter bee escape.

Finally partly open the entrance block, and then leave them alone for three days. Watch them flying in and out, but don't interfere.

Out in the cold?

Some bees may still be crawling around the inside of the nuc box, and it is possible that the all important queen is amongst them.

If this is the case, brush or knock – a sharp knock with the hand – them down into the brood box before putting on the glass quilt. A few puffs of smoke may be necessary to push them down the frames.

A few may still be left. Ensure that the queen is not amongst them. She should be easily recognisable by her larger abdomen and shorter wings, and must be carefully transferred into the brood box if she is. Simply pick her up by a wing between a delicate forefinger and thumb.

Then prop up the nuc box at the hive entrance so the bees can run up the landing board and join the queen inside. They always follow her. Do the same with the skep, and all the bees will soon find their way into the hive.

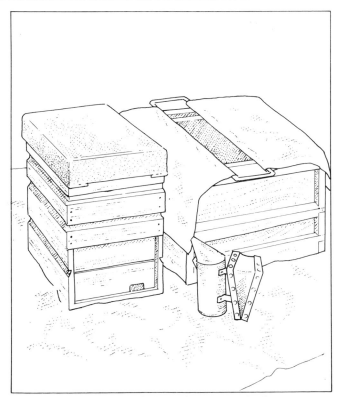

Have the manipulation cloth ready to cover the frames as they are transferred one by one to the brood box.

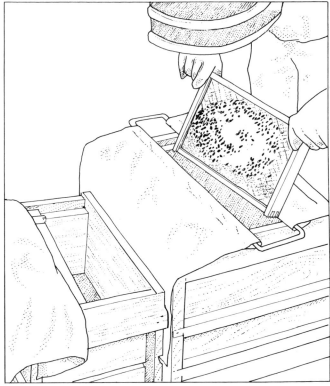

A soft cloth should suffice to cover the nuc as you do this. Ensure the frames go in the way they came out.

Feeding the Bees

After hiving the bees should be left for three days. This gives them time to settle down, organise themselves to fight off any robbing by other bees, and find their own food. During this period you can remove the roof and watch them through the glass quilt, but don't interfere with the frames.

Spring Feed

The bees' first task is to draw out the new wax on your new frames to give the queen more space to lay. If for some reason they won't or can't do this, their little minds will turn to swarming which is the last thing you want at this stage of your beekeeping career.

The bees must be stimulated to draw out the wax, and to do this you give them a first spring feed once the three days have passed. The spring feed is made up of 1lb (0.45kg) sugar dissolved in 1 pint (0.56 litre) warm water, fed in a quarter gallon (1.13 litre) plastic feeder. It is enough to stimulate them to draw out wax, but not so much that they don't bother to go out and find their own food.

Feeding must be done in the evening when bees are tucked up in their hives. It needs great care, for if you drip any sugar syrup around the hive it will attract 'robber bees' the next day which will then plague that hive.

There is a simple technique to avoid dripping:

1 Carry the feeder down to the hive in a large bowl or similar receptacle.

2 In one rapid movement invert the feeder over the bowl so that any drips fall into the bowl rather than on or around the hive.

3 With the roof removed place the inverted feeder over the open Porter bee escape. On a National you should leave a slight opening (one bee space) so that any errant bees can get back down. On a WBC it is not necessary, since errant bees can fly back down to the entrance inside the lifts within the hive.

Great care must be taken not to drip, or bees will start 'robbing'. Carry the quarter gallon feeder down to the hive together with a large bowl.

Remove the roof and place the bowl on the glass quilt. Quickly invert the feeder bucket over it – any excess syrup will pour out and not splash.

Further feeding

The object of this feeding is purely to get the bees to draw out the wax. The feed stimulates the wax glands of the wax making workers, and like little robots they go to work.

When the feeder is empty refill it and continue to supply sugar syrup until the bees are working on the insides of the two end frames. Depending on general weather conditions and what the flying bees are bringing in from outside, your bees could reach this stage in a week.

You should keep a record every time you work on the bees. The easiest way is to jot notes onto a hard card (bees eat paper!) which is left on the glass quilt under the roof; it can double as a handy quick cover for the open feed hole.

Your progress should be along these lines:

1st Sunday. Hive bees.

2nd Sunday. Two ¼ gallon (1.13 litre) buckets fed so far; bees getting near to the ends of the frames. Progress will of course depend on the weather, for if it's been raining every day the sugar syrup will have been their only food and will have been diverted to other uses besides wax making.

3rd Sunday. Bees almost covering outside frames. Time to look through them, and turn outer frames so they can work on the remaining two undrawn sides.

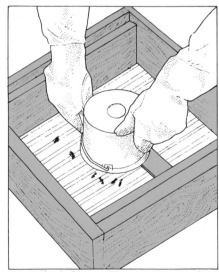

An airlock forms. Place the inverted feeder over the central feed hole – the bee escape must obviously be removed.

Keep a note of the state of your bees. This is particularly important when you have more than one hive, and it becomes impossible to remember what you did last and when. The 'note pad' must be bee proof – plastic or hardboard is fine, but they will eat cardboard. It also acts as a handy quick cover for the feed hole when the bee escape is not in position. You need to record the state of the frames, when the hive was spring cleaned, when new frames were added, when the queen was marked, when supers were added and taken off, etc. The more complete the record the more efficient your beekeeping will become, especially if you have more than one hive.

Going Through a Hive

When it's time to go through the frames of a hive to see how your colony is faring, choose a fine, warm day. It's not something you want to do often, because no matter how fascinating the bees may be, every time you open the hive and start pulling out frames you set the colony back some 24 hours.

The best time to go through the hive is around 1pm when many of the worker bees will be out foraging; they are the ones who make the most fuss, and also the ones who sting. Make sure you are properly clothed from head to foot, and have your smoker going with hive tool to hand. At this stage of beekeeping, it is much easier if there are two of you.

You will also need your white sheet. Bees that fall on it can easily be seen, but searching through grass for your precious queen would be a nightmare!

Start by placing the sheet some 10 feet (3 metres) from the hive. Then give the hive entrance a gentle smoke and wait for three minutes. Take off the roof and put in the Porter bee escape. If it is a WBC take off the lifts; if a National take off the empty super on the top.

Lift the floor and brood box together, and carry it over to the sheet, facing the entrance a different way. This is the first two person job! Leave it for a further 15 minutes, allowing flying bees plenty of time to get clear.

By the time you return the flying bees should have gone, leaving a hive full of harmless baby bees. Prise off the glass quilt with the flat end of the yellow hive tool, giving it a twist to break clear of the frames as you lift it. Bees will be clinging to its surface, so carry it to the site of the hive, placing it besides an upturned skep which is acting as temporary home to the flying bees.

At the moment the glass quilt is removed, get your companion to drop the manipulation cloth over the exposed frames before the bees come up to see what is happening. With careful handling you should not need to smoke

them; when smoke is used unnecessarily it will agitate the bees. However you should keep the smoker on the ground near at hand.

You are now ready to start working through the frames from one side of the box to the other.

The beekeeper above is working at the hive site, and will be plagued by the flying bees – always move away so you can work without angry buzzing! He has also left the brood frames exposed – always cover them over.

GOING THROUGH A HIVE

It is much easier to move the hive a short distance to get rid of the flying bees who will return to the original site. The roof and any supers can be removed for carrying.

If the glass quilt is propolised, you will need the flat wedge end of your hive tool to free it. If the frames lift with it, twist the quilt until it breaks clear from them.

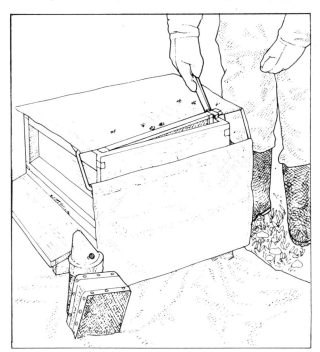

The frames should be covered with the manipulation cloth for inspection, used to expose one frame. The hooked end of a red hive tool is good for frame lifting.

Little smoke should be needed if you are working away from the flying bees. Keep the manipulation cloth down over the frames to avoid disturbing the remaining bees.

Turning the End Frames

When looking through frames, start at one end of the box and work through to the other. Use your manipulation cloth to cover all the frames except the one you are involved with, which will keep the bees happy. Prise each frame loose as you did in the nuc box. The easiest way is to hook the ends free with the hooked end of the red hive tool until you can grasp them with your fingers.

To give yourself room, take the nearest frame out, examine it, and place it by the side of the brood box while you go through the other frames.

Frames only need to be a little bit drawn out to attract bees. When the bees reach the end frames they will start to draw them out, building up the depth of the wax foundation with their own wax. However you want them to get to work on the other side, and all they can can see is a brick wall. So to get them to draw out both sides, put the undrawn sides back facing inwards with the metal ends bent out to suit. This is one instance when you can turn frames round. In a matter of a few weeks the brood will probably not have spread this far from the centre of the hive; if there were any brood the frames would have to go back in the way they came out, with the brood fitting together to keep itself up to the required temperature for hatching baby bees.

If the bees have not got to the stage of starting to draw out the end frames, leave them in the same way round until they have. However now the hive is open, you have a good opportuanity to take a first look through the frames. It's much better to do it early in the season, rather than later when the hive is going flat out.

Examining frames

When examining a frame you must never hold it flat, or the bees can quite simply fall of, Always hold a frame vertical, and keep it vertical while you turn it to examine the other side. To do

THE MANIPULATION CLOTH
Careful use of the manipulation cloth allows you to expose and examine one frame at a time so that the bees stay down in the box. The cloth is heavy enough to stay in place if there is any breeze.

You also need a 'soft cloth' when working on a hive. Any soft cloth will do, cut to a square which slightly overlaps the hive boxes. The soft cloth keeps the bees down while super or queen excluder or clearing board is put in position. It can then be carefully and slowly pulled out.

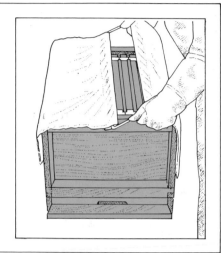

EXAMINING FRAMES

 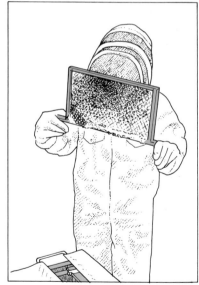

Never hold a frame flat – bees and honey can drop off it. Examine one side of the frame, and then turn it to examine the other by dropping your

left hand (left pic), rotating the frame through 180 degrees anti-clockwise about the top bar, and dropping your right hand (right pic).

this you need a simple technique:
1 Drop your left hand until it is immediately below your right hand.
2 Rotate the frame through 180 degrees to the left.

3 Drop your right hand until it is level with your left hand. You can then examine the other side of the frame, and reverse the procedure to get it the right way up.

What to Look For/Time to Add a Super

What are we looking for when we go through the frames?

● Drawn out new wax which shows that the wax making bees are working.

● Honey stored in the drawn-out frames, sealed with wax and recognisable by its light colour.

● Pollen stored in the drawn-out frames.

● New laid eggs and larvae ('unsealed brood') – little white things in the cells of the drawn-out frames.

● Sealed brood (the next stage on from unsealed brood) in the drawn-out frames, recognisable by its dark colour.

● The queen.

You should see all of these things in a healthy, growing colony. The bees should seem to be going about in a purposeful manner, with workers returning with brightly coloured back legs laden with pollen.

If you don't see the queen, don't worry as it's easy to miss her amongst the mass of bees. If you don't see newly laid eggs on newly drawn-out wax, do worry – the bees may have come to a standstill because the weather has been too poor for foraging and they need feeding, so give them some more spring sugar syrup.

Putting the hive back

If the bees haven't started on the end frames, you should wait until they do before adding a super. If it's a poor summer they may not get to this stage until the second year.

Ensure that all the frames are correctly back in the brood box (use coloured drawing pins on one side of the frames if you are unsure about getting them back the right way round); cover it with the manipulation cloth; carry it back to the hive site moving the skep aside; replace the manipulation cloth with the soft cloth and then the glass quilt; and replace lifts (WBC), empty super (National), and roof. Both feed hole and entrance should be fully open.

Time to add a super?

If the bees have started drawing out the end frames it won't be long before they and their queen are pressed for space in the brood box. When putting the hive back together it is the ideal time to add your first super.

The super must be to hand on the white cloth, complete with its frames. When you've replaced the brood box frames, quickly place the manipulation cloth over the empty super and replace it with a soft cloth to keep the bees down in the brood box.

Place the super on top of the brood box and whip the soft cloth away. Then return to the hive site, and reassemble. You will need another lift for a WBC, and another empty super for a National.

You may find carrying the brood box and super together is too heavy. If so you can take the brood box to the hive site first, and then install the super in situ. If the bees are aggravated, leave putting on the super for half an hour to give them some time to calm down.

With a super on it is worth feeding the hive more sugar syrup to encourage the bees to get to work drawing out the frames. Feed them that same evening with ¼ gallon (1.13 litre) to encourage them to make their way up to it; but don't give them more or they will give it back to you in the honey.

Your bees are now housed in a 'brood and a half' – a brood box and half a brood box. The half box is exactly the same as a honey super, and come autumn we will advise you to take it off the hive so the bees can endure the winter in the comfort of the main brood box.

ADDING A SUPER

Having examined the brood frames, bees are covered over by a soft cloth. This means of keeping them down becomes tricky if there is any wind!

Place the empty brood super on top with the glass quilt acting as its cover, and carefully draw out the soft cloth before reassembling.

An emerging new-born queen

Above: You can clearly seen developing larvae (1), sealed brood (2), a little sealed honey (3), and three queen cells (4) growing like great acorns on this brood frame.

Left: Workers tend to day old eggs (5), larvae and sealed brood in various stages of development. Pollen which is recognisable as dull coloured splodges in cells near the brood can be clearly seen in the photo on page 81.

Right: The marked queen (6) is at the centre of her cohorts, working over the dark brood area of a healthy brood frame. Note the lighter patch which is sealed honey for stores on the perimeter of the brood. The brace comb (7) built out from the top and side bar is fairly normal, and should be scraped away. Correct spacing of the frames ensures the bees do not waste their efforts building it.

Bee Stings/Bees and The Neighbours

The 7th segment of a bee is its sting. The sting is barbed, and it catches so effectively in human skin that it disembowels the bee which will then die. The bee doesn't have this problem when it stings other bees or small enemies such as mice. Having done its work the sting withdraws easily, and the bee can continue on its way.

Despite the fact that it has been torn from the bee, the poisonous sac which supplies the sting will continue pumping the alkaline venom into a human for around three minutes. Therefore the quicker you get the sting out of your skin the less severe the effect is likely to be, and the technique for doing this is clearly shown in the accompanying illustrations.

For those who have still to be stung the pain is nothing to worry about, though the effect is heightened if the sting goes straight into a blood vessel. The small area of red swelling which usually marks a sting can become an uncomfortable swelling which lasts a few days – areas which are vulnerable in this respect are ankles and wrists, so never look into a hive with them unprotected. You should also always take care to protect your face as a sting on the throat or mouth can lead to serious swelling, possibly closing the air passage and requiring a visit to the hospital. Stings to the eyelids should be taken equally seriously.

Antidotes

The best antidote is the natural immunity which many beekeepers build up against bee stings, to the point where they hardly notice them. However this immunity can disappear, sometimes influenced by the introduction of a new medication such as steroids for arthritis which in a few cases can induce a serious reaction.

Bee shops sell 'bee sting antidotes' and some recommend rubbing antihistamine cream on the swelling, but all agree that these remedies are only

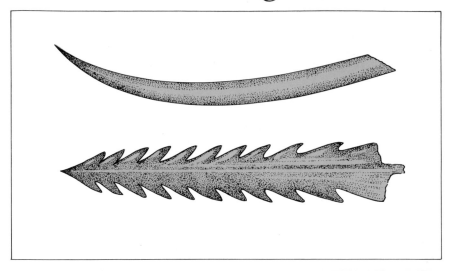

Above: BEE STINGS COMPARED: The queen has a curved sting like a scimitar; the workers have a defensive barbed sting which gets lodged in human skin. The drones have no sting at all.

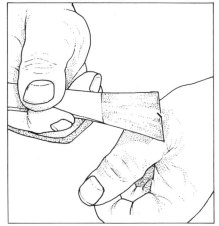

Right: A hive tool comes in useful for removing a bee sting, or the thumb nail can be used.

Below: THE RECOVERY POSITION. If stung in the mouth or throat, breathing may become difficult. The first aid 'recovery position' supports the patient with the airway open while help is called.

partially effective. Some beekeepers resort to a cold compress composed of honey, iodine, onion liquid or lemon juice placed over the swelling, but the result is almost certainly more mental than physical and as such most helpful to children. If a baby is stung the cold compress can also work well, and as long as the baby behaves normally and does not become feverish you have no need to worry.

A very few people are unfortunately hyper sensitive to bee stings to the extent that they may become unconscious a few minutes after being stung. No time should be lost in calling for a doctor or ambulance, and in the meantime they should be laid down with their airway unobstructed (a technique that will be familiar to first-aiders), and kept warm with their clothing loosened.

If someone is stung in the mouth or throat, the authorised First Aid Manual recommends giving them ice to suck to reduce the swelling or washing their mouth out with a solution of a teaspoon of bicarbonate of soda in a tumbler or water. If their breathing becomes difficult they should be placed in the first aid recovery position. An ambulance should be called to stay on the safe side.

Some beekeepers know that they will react this badly to stings, and for them the most immediate and effective treatment is an aerosol bronchodilator spray which gives an immediate boost of andrenaline. Matters of this kind must of course be referred to your doctor, who should be able to advise on potential problems.

Scare stories
It is generally reckoned that 200 stings could kill a man and rather less a child – the human body cannot take that kind of abuse. Bees will never attack in such numbers unless they themselves feel they are under major attack. Instances could be:
● You're carrying a hive, and drop it.
● You're transporting the bees by car and brake so hard that the hive falls over and breaks open.

Obviously both these situations can be classified as 'serious handling' for

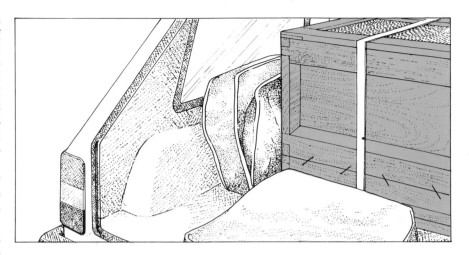

which you must be properly clothed, and therefore safe from stings. Even attacks by the Africanised 'killer bees' (page 168) which are dominant in South America have only happened when the bees have been disturbed.

Bees & the Neighbours
You owe it to your neighbours not to make a nuisance of your bees. They don't want to be stung, and there is no reason why they should be if you manage your bees sensibly.

Some beekeepers like to go through their hives every 10 days, searching for queen cells and squashing them as they find them. This is a technique used to prevent swarming which is covered in the second year section of this book. It is ok if your hive is in a wide open space in the middle of the countryside; it is extremely foolhardy if you live in a densely populated area or have close neighbours. You may be fine fully clothed in your beekeeping gear, but the neighbours won't be happy when they're visited by bees being maddened by your attack on their hive.

So before looking at your bees, THINK OF OTHERS AND DON'T BE SELFISH! Otherwise you may have a riot on your hands . . .

Simple rules
There are simple rules to avoid trouble with close neighbours. The last thing you want if for them to label you a 'menace' and complain to the District

If you travel with a hive, wedge it securely with frames 'fore and aft'. Note the staples holding the boxes, strap, and ventilation screen.

Council:

1 Never place a hive next to a footpath.
2 Two hives are about right for a small garden around 30×50ft (9.1m×15.2m). If you want more hives, you should keep them elsewhere.
3 If you want to go through the hive you should tell your immediate neighbours. If they want to join in encourage them to give a hand; if they are scared gently suggest that they stay indoors.
4 Only go through one hive at a time. It is of course much quicker to go through as many as you have all in one go, but hive 1 tells hive 2 that they are being interfered with, and by the time you get to hive 2 you have some very bad tempered bees.
5 Always use the technique of moving the hive a short distance from its site so that you cause minimum fuss with the flying bees.
6 Going through your hive at mid-day every 10 days during the summer is in essence a good idea. In practice it is a very bad idea if there are neighbours around.
7 If 3–6 are a problem, transport your bees somewhere else to go through them.
8 At harvest time reward your neighbours' patience with pots of honey!

The First Year:
A Thriving Colony

If all has gone well, your bee colony could grow to filling a brood and a half by June. To get this far so fast you would need a strong nucleus installed in April, and consistent good weather. If everything continues well you could add a super and harvest as much as 25lbs of honey by the end of the season – but this would be a very good result when the bees have been so busy drawing out new wax in their first year.

Arthur Chitty, one of many veteran beekeepers, waits for a swarm to settle at his apiary. Swarming is a natural instinct of honeybees which the beekeeper attempts to control. It takes place in the early months of summer, but unless you neglect your bees and leave them overcrowded is most unlikely to happen in your first year. You must give your bees more space as they require it – they owe no allegiance to an incompetent beekeeper.

When to Add Honey Supers

The bees need to draw out the wax because it's their natural home for pollen, honey, eggs, etc. In a natural state they would create their own wax home from nothing, settling in somewhere suitable such as a tree or roof. Initially they would make cells for worker bees, and then make larger drone cells round the outside which require less wax for more honey storage.

In the controlled envioronment of a hive the beekeeper gives them the basic foundation of preformed worker cells, as part of his plan to encourage the queen to lay and produce workers – more workers equal more honey.

The queen lays according to how much she is fed. As the season progresses she lays faster as the nectar comes in faster. If she runs out of space to lay the workers also run out of space to store food. They have to store the nectar in their stomachs which automatically triggers their wax glands, encouraging them to get to work building out cells on the new, undrawn frames.

The time to add supers is when the bees have run out of space on the frames – there are just too many of them to congregate on the space provided. You have already increased your brood size to a brood and a half; when the bees are clustered on the end frames on that half they need more space. Turn the end frames to make them draw out both sides as you did with the brood box, and then add a super which will be for honey storage only.

Adding the super
As with many beekeeping operations, two pairs of hand make it easier than one. At least one can read out the instructions from the book!

1 Prepare your super with a full set of frames. Remember that you will need an extra empty super on a National, and an extra lift on a WBC.

Remove the roof of the hive. Free the glass quilt with the hive tool, and then use a soft cloth to remove it.

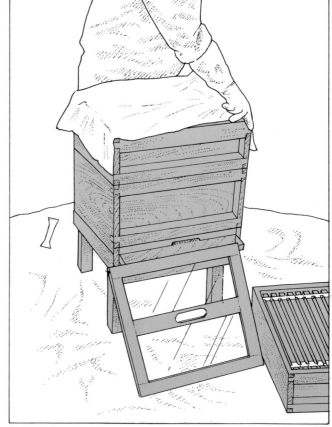

Put on the manipulation cloth and have the new super to hand (bottom right). The glass quilt acts as a ramp.

2 Time the operation for the middle of the day. It's a quick job; you won't be interfering with the main brood box; and there is no need to move the hive.

3 Get fully dressed with your smoker going. However there is no need to smoke unless the bees become aggressive.

4 Take the roof off the hive. Lay your soft cloth across your shoulder and put in the Porter bee escape. Lever the glass quilt at all four corners; twist and lift; and at the same time drop the cloth over the uncovered frames.

5 Stand the glass quilt as a ramp up to the entrance for any bees that are left on it. Cover the frames with the heavier manipulation cloth, and remove the soft cloth.

6 Expose the second or third frame in from either side of the brood super.

7 Take out that frame, and examine to make sure there are no eggs or brood on it. It is extremely unlikely that the queen has got so far, but if she has, put it back and try the next one outwards. You are looking for a drawn out frame which may possibly have some honey, but has not been visited by the queen. As long as the bees have started to draw it out that is all you need – they leave a smell which will attract them back to it in the new super.

When you find a suitable frame take it out complete with bees, replace it with a new frame, and then repeat the operation on the other side of the hive.

If the bees have not started to draw-out the outer frames (second or third in) on either side, they still have a long way to go and it is too early to add a super.

8 Put the two drawn-out frames together in the middle of the new super. You can leave the bees on them, but check to make sure the queen is not amongst them.

9 Take the manipulation cloth off the top of the brood and a half. Place it to one side, using your soft cloth to cover the brood and a half. Use a little smoke if necessary.

10 Put the queen excluder on the brood and a half, the new super on top of that, and then remove the soft cloth. The new super must be covered with the glass quilt, ensuring the feed hole is open. Reassemble the hive with additional lift (WBC), empty super (National), and roof.

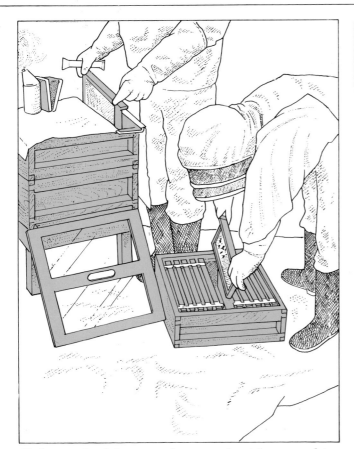

Find a couple of drawn out frames, and put them complete with bees in the centre of the new super as shown.

Keeping the bees down with the soft cloth, put the queen excluder on the brood; then the super, glass quilt, etc.

The Queen Excluder

The queen excluder is like a bee's cattle grid. The workers can pass between the bars, but the larger queen and drones cannot and are therefore kept down in the brood box.

It is possible that some drones may have landed up above the queen excluder when you moved the two frames. They can't get down if it is a single wall hive, so be sure to take out the two frames after a couple of days and leave them by the side of the hive for a few minutes. If any drones are on them, they will fly free. Alternatively leave the frames in place with the glass quilt off the super for around 10 minutes so that any drones have time to fly up, out and away. In a WBC they should find the way out without help.

Why?

Why do we have a queen excluder? Because we don't want eggs, baby bees and other bits and pieces mixed up with our honey, and we want to get rid of the workers when we remove it. Commercial beekeepers mostly don't favour queen excluders, but for the less experienced they make beekeeping much easier.

There is an old kind of zinc queen excluder which should be avoided. It is a flat sheet with cut-out holes that tear the wings of the bees as they pass through, and it also warps. The kind to go for is a grid made up of wire rods, but the workers still don't like to pass through because they find them cold! Even though they need space for their stores they may refuse to go up unless there is something to attract them, and that is why you must have a couple of frames which they have started to draw out and will be enought to lure them upwards.

NB The queen excluder must be removed when preparing for winter. If left on you can get a situation where all the stores are above the queen excluder, while the bees are starving below because they have to stay in their winter cluster round the queen. This is covered in the 'Wintering Down' section, starting on page 84.

Workers come up through the queen excluder to store honey in the supers above. The fatter queen stays below.

How Bees Forage

Approximately half of the workers in a hive have domestic duties. The rest are foragers, which is the most debilitating part of a worker's life inevitably leading to death from overwork!

At the height of the season workers end their domestic duties at approximately three weeks old, and then last about a further three weeks foraging. If you look closely you can see the 'old ones' landing with their wings all torn from their hardships, trying to crawl back into the hive.

A very small number of workers are nominated as scout bees who fly out and find the honey and pollen sources where the foragers should go. These sources are continually changing as the season progresses and different crops come into flower. While one party of foragers may be hard at work on a pear orchard, others may be out of work having exhausted another source. Much of their time is spent waiting for the return of scout bees who will tell them where to go next.

The bee dance

When a scout bee returns, it picks up whatever out of work foragers are there, and directs them where to go with a 'dance' in the dark of the hive, demonstrating the distance of the source and where it is in relation to the sun. The scout also regurgitates some of the nectar it has sucked up to give the foragers the taste, and then they head off looking for it. With pollen the foragers lick the scout bee to get the taste of it.

There are three bee dances which the bees 'feel' with their antennae, for with their poor eyesight they certainly can't see what's going on. The first bee dance gradually merges into the third as the distance for forage increases.

The round dance is used to indicate forage up to a distance of 80 feet (25 metres), but is not specific as to direction or distance. It consists of a circular dance with reversals of direction at intervals. A complete circle may be danced a variable number of times, followed by a complete change of direction.

If the distance of the forage is 80–325 feet (25–100 metres), the nature of the

At the height of the season the landing board becomes like Piccadilly Circus as the foragers come and go!

dance changes. It evolves into the more detailed wagtail dance in which the dancer indicates the direction, distance and quality of the nectar sources by wagging her abdomen from side to side and dancing up the middle of the circle. If the distance is over 325 feet (100 metres), the wagtail dance alone is used to tell the bees where to go.

While dancing the scout bee will regurgitate nectar and stop frequently to offer it to the foragers who will use the odour as the final clue to finding the source. If the scout bee only carries pollen, it will offer it around in much the same way with the foragers 'watching' (with their antennae) much more closely.

The flight path

Once they have been told where to go, the foragers come out and take off in a circular orientating flight around the hive before heading off in their chosen direction. This also allows them the opportunity of discarding husks from stored pollen.

However once they know the way they will usually head straight off, only returning to the orientation circles when they are going to a new source.

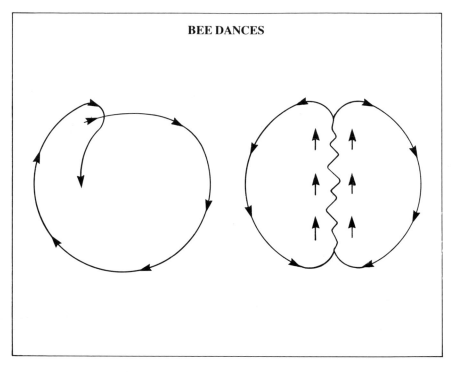

BEE DANCES

Above: The 'Round Dance' (left) indicates forage up to 25 metres; for longer distances the dance takes on the characteristics of the 'Wagtail Dance' (right), indicating direction, distance and quality of nectar.

Below: How the foragers find the way. The scout bee dances to show them where and how far; they leave on a first orientation flight using the sun for navigation; they find the source; and then fly back and forth direct.

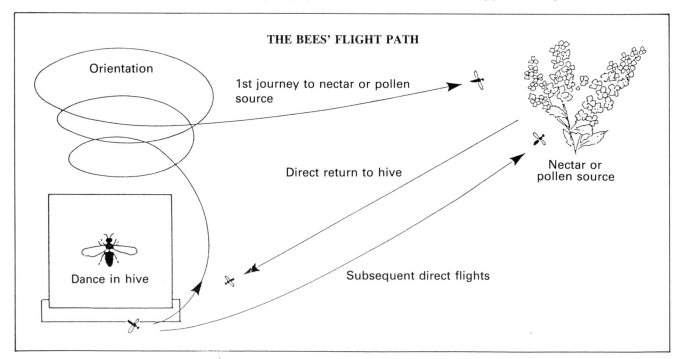

THE BEES' FLIGHT PATH

Orientation

1st journey to nectar or pollen source

Direct return to hive

Nectar or pollen source

Dance in hive

Subsequent direct flights

What the Bees Collect

Foragers will fly up to three miles (4.8km), but the general rule is the nearer the better. If the source is oil seed rape they will fly up to five miles (8km), such is the return for their labours. (The oil seed rape honey flow is dealt with on page 106). They will not leave a pollen or nectar source until it is finished, and even when it has the scout bees will return periodically to check that it has not started up again as is sometimes the case.

Nectar

Foraging only takes place during the daylight hours when nectar rises with the heat of the land. Nectar is basically glucose, fructose and water secreted through the flowers of certain plants in different concentrations.

To collect nectar the bee extends its tongue down into the flower and sucks the nectar down into its honey sac. The total amount it can carry is miniscule – from 20 – 100 milligrams which is about enough to cover the head of a pin!

The bee then returns to the hive and regurgitates the nectar into the mouth of a hive worker, who in turn takes it to an empty honey cell where it is pasted around the walls so that the water content can evaporate. At the same time enzymes from the bees' stomachs are working on the nectar, causing the fructose and glucose to invert which leads to the formation of honey. When the water content is down to a maximum of around 17%, all of the honey which is ready is then collected into one cell which is capped over for storage with new wax when full. The bees produce honey for their own food; the beekeeper keeps bees to farm that honey, feeding the bees with alternative food when necessary.

If the water content is 20% or higher the bees will not cap it over. Never spin off a honey frame (except oil seed rape) unless at least three-quarters of the honey cells are waxed over. If they are not the honey has too much water, and will in time ferment and turn to alcohol.

Pollen

Bees may collect nectar, pollen, or both. The bee dance tells them what they can collect; the state of the hive tells them what they need.

To collect pollen a bee dives down into the flower and covers herself in it. She then hovers while scraping it into the pollen baskets on her back legs, setting up an involuntary fertilisation (pollination) chain when she flies to the next plant. Bumble bees are well know for the useful role they play in this.

The bee then takes it back into the hive, and scrapes it off her legs into a pollen storage cell. These are situated round the brood, acting as protein food to make the exoskeletons (the outer skins of the bees) of the larvae develop.

You can see the bees flying in with their pollen laden legs, and the colours – orange, yellow, etc – give an indication of the pollen sources. Sometimes they bring in too much pollen for the hive, in which case it is rejected and the pollen gatherers join the out of work bees.

MAIN NECTAR SOURCES

MAR APR MAY JUNE JULY AUG SEPT OCT

Brassicas – including oil seed rape. Autumn sown crop giving largest yield.

Dandelion – all year round but best nectar yield in spring.

Sainfoin – ideal June gap plant.

Sycamore – a very dark but highly flavoured honey.

Clover – white clover best – after a heavy dew or rain nectar rises.

Lime – follows clover – lasts ten days.

Blackberry – grows on any soil – flowers until frost arrives.

Willow Herb – very good plant secretes large amount of nectar

Heather – gives a large crop if conditions right.

POLLEN SOURCES

	Early	Middle	Late
COLTSFOOT	●		
ANENOME	●		
WINTER ACONITE	●		
DANDELION	●	●	●
ELM No Nectar	●		
HAZEL No Nectar	●		
BROOM	●	●	●
GORSE		●	
MEADOWSWEET		●	
BLACKBERRY		●	●
IVY			●
SWEET CHESTNUT		●	
CLOVER		●	
MICHAELMAS DAISIES			●
HEATHER			●
LING			●
BRASSICAS	●	●	
CHRISTMAS ROSE			●
SYCAMORE	●		
HORSE CHESTNUT	●		
HAWTHORNE		●	
MAPLE	●		

Foragers fit in with the main pollen sources (left) and return to the hive with brightly coloured 'baskets' strapped to their hind legs (above). It becomes the main brood food of the colony.

Water and Propolis

Bees may also carry water, and rely on a bee dance to tell them where to go. Since water is comparatively odourless, the scout bee will use its Nasonov gland to emit an odour at the source to guide in the water carriers. They may also collect propolis – see page 164.

Right: Honeybees are not the only foragers, though they are certainly the most efficient. The humble bumble bee lives in small colonies, and manages to collect just enough for its needs and no more. Here a bumble bee visits sugar supersaturated oil seed rape, an equal favourite with the honeybee.

Taking off Honey

You can take off some honey if your hive has reached a brood and a half, preferably plus a super. Never take honey from the brood box. The bees need it more than you.

Brood and a half plus super

In your first year the hive will have become established much too late for the oil seed rape honey flow which is completely out of step with the main season. That is a second year subject, dealt with on page 106.

The end of the principal UK flowering season will be late August which is the time to take off your honey. After this the bees should be allowed to keep any late season honey which they harvest.

Get yourself a wooden clearing board (crown board) with two Porter bee escapes. Make sure they are in the right side, so that the bees can go down but can't come back up.

It is a time of year when there are a lot of workers with little to do, and consequently the risk of robbers moving in from other hives is high. You may inadvertently shake some honey onto the ground, so only work on the hive in the evening when all bees are tucked up for the night. Close the hive entrance to a few bee spaces to make it easy for your bees to defend and make sure there are no holes or gaps in the hive. With a WBC, plug the round holes in the roof's ventilation cones with a couple of matches.

Is there any honey is there? Get well dressed and only remove it late in the evening.

Operation 1

On the first evening your job is to put on the clearing board. The bees live in the brood and only visit the super to deposit nectar or draw out wax, or because they need more space. With the clearing board in place they can go down out of the super, but they can't get back up into it. Two pairs of hands will make this operation easier:

1 Move gently and quietly. You must be fully clothed, but don't use smoke unless you have to. Remove the roof, and then lifts (WBC) or empty box super (National).

Have your soft cloth ready, and lever and twist off the honey super. Place it across the upturned roof, at right angles so you squash the minimum of bees. Throw your soft cloth across the exposed queen excluder which now sits on top of the hive. This will keep down any bees.

2 This is the stage when the queen excluder should be removed and replaced by the clearing board. Lever it off with your hive tool with the soft cloth still covering it. As it comes free pull the soft cloth off and once more throw it over the exposed frames. Then place the clearing board on top of the hive, removing the cloth when it is in place.

From now until the spring the clearing board (also called 'crown board' or 'wooden quilt') replaces the glass quilt as the top cover of the hive. The evenings are getting cold, and unlike glass its wood is condensation free.

If you have a spare super and frames, a slightly different method can be used. Leave the queen excluder in position (but remove it when wintering down), place your spare super on top, then the clearing board, then your full super. There are a lot of out of work bees around at this time of year, and the spare super gives them room without crowding out the brood and a half. They may also start to draw out the wax in the new frames.)

3 Replace the honey super on top of the clearing board, trying not to squash any bees. Leave the hive for 24 hours.

OPERATION 1: Lever and twist the honey super free, temporarily placing it on the upturned roof with a cloth over the brood.

Remove the queen excluder by levering it free, and carefully and gently drawing it out from under the soft cloth which will keep the bees down.

Place the clearing board over the soft cloth with bee escapes in position. Remove the cloth; replace the honey super; and reassemble the hive.

Operation 2

Return 24 hours later as it becomes dusk, once again fully clothed. If for any reason the weather is unsuitable – wet or windy – leave the hive a further 24 hours.

1 Remove the roof of the hive, plus lifts (WBC) or empty super (National). Check through the glass quilt that all the bees have vacated your honey super; if not, leave for another day.

2 Standing behind the hive, lift the honey super off, and quickly take it to a closed room which you must keep bee proof. You are stealing the bees' honey, and they will come and get you if they can find out what you are up to or where it is.

Great care must be taken not to drop honey. A sensible precaution is to stand everything on a plastic sheet or sacks laid round the hive, and carry the super on a flat tray so that you take no chances of leaving honey drips to guide avenging bees. If you think you have dropped honey, water the ground to get rid of any smell.

If there are two of you, your helper should reassemble the remaining components of the hive while you carry off the honey super. If you are doing it alone, return straightaway do do this.

3 Go back to the closed room and look through the frames. In your first year the chances of finding an appreciable amount of honey in this super are not high.

Remove the frames with honey, but only extract it for your own use if at least three-quarters of the honey cells are sealed. If they are not, either put the frames back on the hive so the bees can continue to work on them and prepare their furture stores, or spin off the unsealed honey and feed it back to them as part of your pre-winter preparation (page 86).

NB. Oil seed rape honey is easiest to extract before the cells are capped over. If it doesn't shake out of the frame, it is ready for spinning-off. If you leave it too long it may solidify, since it granulates much more rapidly than other honeys.

OPERATION 2: First check through the glass quilt that the bees have vacated the honey super. A few bees left can easily be brushed off the frames.

Remove the honey super and quickly steal away, taking great care not to drip as you go. This operation must be performed at dusk with no bees flying.

Once in a closed room, you can look through the frames. The honey should be mostly capped-off; if not it is likely to be too unripe to keep.

Brood and a half – no super

It is really better to leave honey in the brood super to the bees. However if your hive has just got to a brood and a half with no honey super, you may feel you must have some honey. Proceed as follows:

1 Go to the hive in the evening, fully clothed and with the smoker lit. Smoke the hive entrance; remove roof etc to expose the glass quilt over the brood box super; and wait the required three minutes. Then move the hive away so you can work on it without being bothered by the flying bees.

2 Remove the glass quilt and replace with the manipulation cloth. Then, working from each side pull up the frames so you can get a quick look at them. If any have sealed honey and nothing else, lift them out and brush the bees into the space left behind. Place them in an empty super for carrying, and replace with new frames – or better still drawn-out frames if you have them. (This would be an ideal place to put back frames with unsealed honey.)

When you get to dark coloured sealed cells which indicate brood, stop and start from the other side of the box. You don't want to interfere with the new born bees, and this is where the all important queen may be. If you take out a brood frame the bees will start to get very angry – the end of the main honey flow is their angriest time of the year.

Take note of the state of the frames you examine, for this will give an indication of how you should winter them.

3 Quickly close up the hive, and take the sealed honey frames to your closed room. This is your harvest which you are going to spin off.

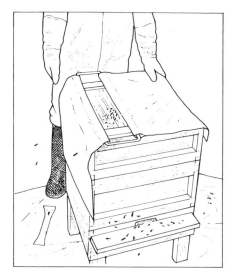

You can take honey from the brood super, though care must be taken to smoke the bees correctly and make sure you are not spinning off brood!

Use the manipulation cloth and a little smoke to check quickly through the frames. You are looking for light coloured honey; stop at dark brood.

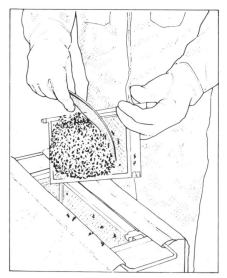

Brush the bees off the honey frames and replace with new frames. You can spin off the brood honey, putting back sufficient stores when winter feeding.

Spinning or Scraping off Honey

You can remove the honey from a frame by scraping or spinning it off. Two things must be stressed:

● It is a sticky business, so stand everything on a plastic sheet.

● The room where you do it must be bee tight, or you will have a lot of very interested visitors.

Spinning off

Spinning off is the neatest and cleanest way, and it preserves the drawn out wax on the frames. You can buy your own hand cranked machine from £100 upwards, or allow from £300 for an electric spinner. Obviously you would need a lot of honey to make spending this kind of money worthwhile, so most hobbyist beekeepers rely on hiring from the local shop or association, or attending an association 'honey spinning evening' when all members can take along their frames and use a communal machine.

Most spinners are simple to use:

1 The machine must be clean to start with; if not wash it out with boiling water.

2 It must also be dry, or the honey will ferment.

3 Most small machines take two frames at a time loaded in cradles. The three types of spin are tangenital, radial, or horizontal.

Before putting in the frames you must run a long, sharp knife over the surface to remove the wax capping both sides. Scrape it into a container or 'uncapping tray' which is a worthwhile extra incorporating a heating element to get all the residual honey out of the wax. The wax can be melted down in an empty tin placed in a saucepan of boiling water. You can use it for your own candle making and other uses (page 160), or sell it back to the shops – the current rate is around £1.15 per 1b (0.45kg) against purchases.

4 Start an electric spinner slowly, or the

This brood frame is full of honey, and darker coloured pollen, but only the white area at the top is capped over for the bees' future stores.

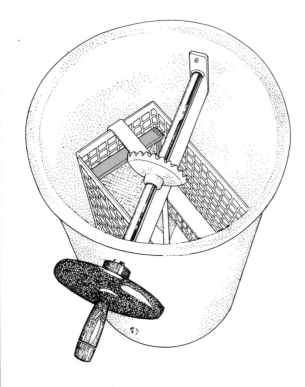

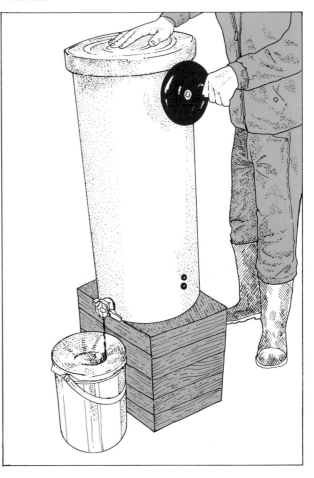

A typical manual extractor will take four uncapped frames dropped into the sides of a cage with metal ends removed. Start spinning slowly; work up to full speed; reverse the frames after a couple of minutes and repeat.

centrifugal force may push the wax out. Note that whether manual or electric, you can only spin off wired frames.

5 The frames will take about five minutes to spin off in an average machine. When no more honey is coming out, store the frames in a super.

6 The frames will be 'wet', meaning they still contain residual honey which you can give back to the bees.

At dusk go down to the hive fully clothed. Remove the bee escapes from the clearing/crown board, and place the super with the wet frames on top. The bees will clean these frames up, taking all the honey down near to their brood for winter storage. The entrance must be closed down to a few bee spaces to guard against robbing.

7 Allow the bees 24 hours to clean the frames. You can then remove them, or leave them until you're ready to winter down the hive if more convenient.

Scraping off/Cutting out

You can simply scrape off the honey with a spoon, but it's a much more primitive method which mixes wax, pollen, and the detritus of dead bees with your honey, and destroys the efforts of the wax making bees.

Simpler still you can cut out unwired 'comb honey' (honey stored in cells built on unwired foundation) with a comb cutter. This is the kind of honey you can buy in a shop like Harrods for a high price, and all the odds and ends mixed up with it are considered by

many to be highly beneficial to your health, particularly any pollen which is stored around the honey.

Some much prefer the chewy taste of cut comb or scraped off honey – others prefer a more refined product which doesn't leave wax in their teeth!

Bottling

If you want clear honey you must filter it through damp muslin which will let the honey run through. Alternatively stainless steel filters are available. You must do this if you wish to sell the honey, and for competitions you will also have to remove all bubbles as they rise.

Bottling for competition is a specialist subject dealt with on page 152. At this stage of the game you should pour what honey you have into clean, dry jars; give thanks; and enjoy what you have gained without worrying about the finer points of presentation.

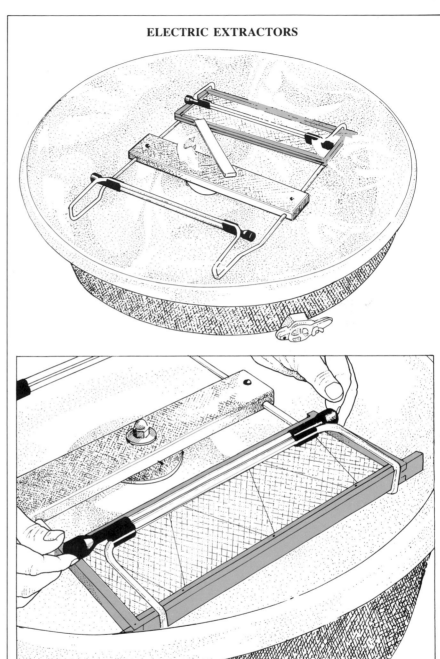

A sharp knife is needed to evenly slice off the honey's wax cappings.

If the wax is wired, you can spin it off. If not, cut out the whole comb.

The unwired comb can then be cut into five pieces stored in 8oz cartons.

An electric extractor which resembles a huge plastic saucer. Two uncapped wired frames are locked in place at both ends of the rotor arm. With the machine plugged into the mains the speed of the revolutions can be adjusted; start off slowly and then build up speed, otherwise you risk spinning the wax completely out of the frames with too much centrifugal force. The main disadvantage of this machine is that it will only spin two frames.

The First Year – Wintering Down

Bees need care if they are to survive the winter without being severely diminished, or in some cases wiped out. You must ensure they have sufficient food throughout the cold season, are well protected from predators, and that the hive is free from damp which is the bees' main winter enemy. Then you can look forward to a healthy hive for year two, raring to go into munificent honey production. . .

A freezing scene! However it's not cold which kills colonies, it's damp. A wintered-down hive must be well ventilated, and protected from mice, woodpeckers and other nuisances. If you have taken away its honey stores these must also be replaced, so the colony is 'fast fed' a large amount of sugar syrup to see it through to the spring. Occasional top-ups with 'candy' may also brighten the lives of your bees. . . .

Feeding in Autumn

Your honey should be spun off by the first week in September. After that any more which is collected should be kept by the bees as part of their winter stores. Ivy is one of the main sources of nectar and pollen this late in the season.

Your task is to prepare them for the winter, and ensure that they survive as a healthy colony. All the honey supers must be removed, and then you must judge whether to leave the colony as a brood and a half or in its brood box alone.

The simple rule of thumb at this stage is that if there is brood in the brood box super you should leave it on; if there is none, but more than 50% honey you should leave it on; if there is no brood and less than 50% honey you should take it off and leave your bees in the warmth and comfort of a single box.

Feeding

Since you have taken away so much of the bees' stores you must feed them in return. What they have left is unlikely to be enough.

Start feeding them in the last week of September. Your aim is to leave them at least 30lb (13.6kg) of stores. They will already have some of their own in the brood box, and for the rest you will feed them with a thick mixture of sugar syrup. Dissolve 16lb (7.2kg) of refined white sugar in 8 pints (4.5litres) of hot water which should be ample for their needs. Brown sugar tends to upset their little tummies.

Fumidol B which is an antibiotic against Nosema can be mixed with the first feed. It is effective for three years, but since it is comparatively cheap we would recommend adding it every season – then you know you are safe.

The easiest way to feed the syrup is with a 'fast feeder', so that they take it down quickly. The fast feeder is basically a super which can hold well over a gallon (4.5 litre) of the syrup at a time. In the middle it has something that looks like a molehill with a hole in

FAST FEEDER

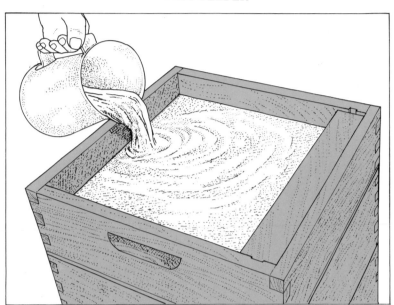

Fast feeder in action. This is the Miller design, which takes approximately 1 gallon of sugar syrup at a time. With exactly the same dimensions as a super it will fit neatly on top of the hive.

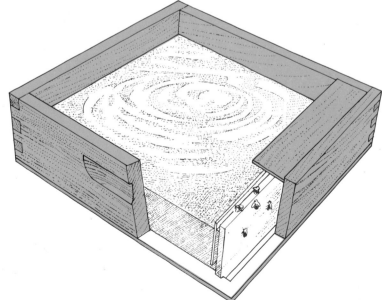

Being fast workers, the bees can empty the syrup in 24 hours. You then refill the feeder until they have been supplied with the necessary amount for winter. A fast feeder is a very worthwhile investment.

the top, or long entrance hole (Miller). The bees come up that hole and take the syrup down into the hive.

Fast feeding
Get fully dressed with your smoker alight, and take the fast feeder down to the hive at dusk when there is least chance of robbing. Close the entrance down to one bee space, and take great care that you spill no syrup. Then:

1 Remove the crown board using your hive tool as necessary; cover the bees with your soft cloth to keep them down; place on the fast feeder; and remove the cloth.

2 Have the syrup standing by in a bucket which must be covered – even the smell gets bees going. Pour in the syrup taking great care not to splash, and when the fast feeder is full place the crown board on top with the feed holes completely covered. This is to prevent the bees finding a way down into the syrup and drowning, and is vital on a WBC and good security on National.

3 Finally place the roof on, and with a WBC make sure the ventilation cones are blocked against robber bees.

The bees can take the syrup down in 24 hours, but if they don't it means they don't need it as yet. One filling should be enough, but repeat the procedure if it is neccessary.

No fast feeder
If you don't have a fast feeder, you can use a 1 gallon (4.5 litres) plastic bucket feeder using the same technique as for spring feeding (page 58). Invert the bucket over a container so there is no chance of dripping round the hive.

This will be slower than a fast feeder but is nevertheless ok. However don't attempt to feed them at this stage with the small quarter gallon (1.13 litre) feeder which will be much too slow and involve you in eight or more visits to the hive.

The idea of the autumn feed is for the bees to take down the syrup into the hive as quickly as possible, starting the process which evaporates the water content so they can cap it over as part of their wax covered stores. If you give them a slow feed they will be fooled into thinking it's the honey flow. The queen will then be stimulated into egg laying at completely the wrong time of year with winter just ahead – they don't want more bees when the hive should be wintering down. Using the fast feeder the flow is much too fast to feed the queen. The bees only have time to store it, avoiding a potentially tricky situation.

Two hives
If you have two hives you must feed both at once. Use a small plastic feeder as a temporary diversion for one, or they will go out and rob the other. Then switch the fast feeder.

The end of feeding
Feeding should be finished by the end of September. If you leave it later and the weather suddenly turns wintery, the temperature may be too low for them to get the moisture content of the sugar syrup down to 17%, and it consequently cannot be capped off. The result is that it will ferment, and be of no use to the bees.

If for any reason you are feeding late, you can feed slightly thicker syrup with less water content. However this can lead to digestive problems for the bees.

Having fed the bees an expert may 'heft' the hive, lifting the back to judge if its weight indicates the necessary amount of stores. Obviously you need to be a highly experienced beekeeper before you can use this method.

A fast feeder in use, being filled with sugar syrup from a watering can – no drips and a perfect aim! Two fillings should suffice for your bees.

Battening Down

Now the hive should be battened down for the winter. Take off the fast feeder and replace it with the crown board, using the soft cloth trick to avoid squashing bees.

When leaving the bees for the winter you have to allow for ventilation. You can leave the feed holes in the crown board open, but the two small openings will tend to suck the cold winter wind through the hive. A better solution is to cover the bee escapes with your note cards, and then insert a match diagonally under the four corners of the crown board. This gives ventilation all round the top with gentle dissipation of the wind, and the space is not sufficient for the bees to get through.

We would also recommend taking out the entrance block completely to prevent the venturi effect of wind funnelling through a small hole. It also helps prevent the build-up of damp inside the hive.

You must put on a zinc mouse guard with round holes to keep out any mice. This is needed when the temperature drops – not too soon as it can scrape the early winter pollen off the foragers' legs. November is the right month to put it on, and the zinc can simply be fastened into the entrance with drawing pins. Remember to press gently; don't use a hammer, or the bees will get very annoyed!

Finally put a strap round the hive to prevent the roof from blowing off, or invest in the optional National 6 inch (15cm) deep roof. Clear any damp vegetation from around the hive, and they will be safely tucked up for the winter.

Storing frames

All the wet frames should have been cleaned by the bees, and are ready for storage. You must clean up the glass quilt using a blow torch and the yellow hive tool as a scraper.

A small butane blow torch is best for the job, though you can also use an

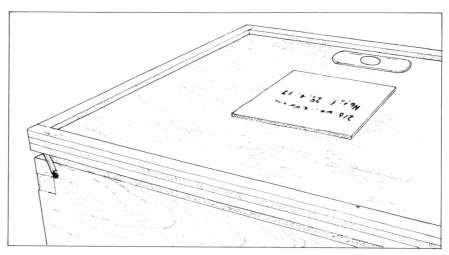

Above: Wintering down. The clearing board has a matchstick under each corner to allow a draught; the bee escape is in place; notes are left on top.

Below: A zinc strip mouseguard is held in place with drawing pins. The bees have plenty of space while tiny mice cannot get through the holes.

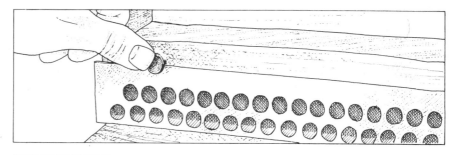

Battened down for winter. The hive has a strap round to prevent any chance of the roof blowing off.

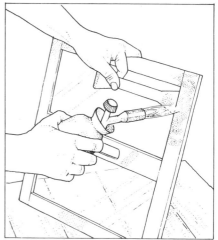

The flame of a blow torch will kill all pests and diseases. Protect the glass with the hive tool as shown.

electric paint stripper. The heat will kill any infection and make propolis and wax easy to remove, but great care must be taken not to break the glass, using the flat end of the hive tool to protect it.

Wax moths

The great winter enemy to your equipment is wax moth. The lesser or more recent greater wax moth likes to lay its eggs in stored frames in the late autumn. It digests the wax on the frames to feed the larvae which hatch out after six weeks. Meanwhile the greater will burrow into the wood of both frames and boxes, living off the castings (old skins) from the bees and feeding the silicone to its babies.

Once in the frames the wax moths will stay there, and when you put them in a hive next spring the weak bees are vulnerable to a full scale take-over. When the frames are full of honey the wax moths will tunnel under the cappings and make it completely unusable. It takes them about 10 days to

destroy a whole honey frame, and that will obviously destroy the hive.

In the same way that we stop moths getting into stored clothes, we can keep them out of stored frames with PDP crystals. Put a tablespoon of the crystals into an envelope, and pin it flap up to the side of each super full of stored frames. The fumes will then permeate throughout.

If you have a second super lay newspaper over the first one, and then place the second on top leaving no gaps round the edges – if there are seal them with brown tape. Do the same again with an envelope containing crystals, and then place the glass quilt on top with the feed hole covered over. Your frames can then be left safely until the spring.

Wax moth likes to breed in warm, dark places, and an alternative method of keeping them out is to store your frames loosely in a covered, cold, draughty open area such as a car port. The wax moth will not breed if conditions stay this way, but if there is a very mild spell you could be in trouble.

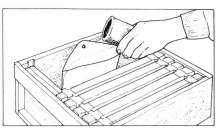

When storing for winter each box of frames should have an open envelope with PDP crystals pinned to the side to keep the wax moth at bay. Put a sheet of newspaper between each super or brood box and leave them stacked with no gaps showing.

If by any chance the wax moth does get a hold, methods of killing it are detailed on page 175. Simple freezing is the quickest and easiest method to rid yourself of this pest, and it can be accomplished with a domestic size freezer.

Below: Overwintered nucs. Snow can confuse bees. They see the bright light and fly out when it's too cold – hang down some sacking to darken the hive.

Mid Winter Care and Candy

Candy is the bees' Christmas present, and is a barometer as to how they are faring. The bees will only take it down into the hive if they need it – if they eat it all they are short of stores, and you must give them some more.

You can buy a ½lb (220gm) block of candy from a beekeeping shop for around 50p, and it is much easier than making your own.

The block should be wrapped in greaseproof paper which the bees can chew off and spit out of the entrance of the hive. They can't do that with plastic. Cut away the paper from the bottom, and place it over an open feed hole with no space round the side. The traditional time to do this is a few days either side of Christmas.

The bees will take the candy down slowly, and will feed on it straightaway for immediate energy rather than storing it. It should take them at least a month to digest a block, but if they keep going through it you should feed more until you are ready for spring cleaning.

Mid Winter Problems

● Woodpeckers will attack a hive and burrow in and eat the bees. A National is particularly vulnerable, and once one starts others will join in. This is usually a problem out in open fields, which can be prevented by covering the hive with wire mesh or black plastic.

● Snow can fool the bees into thinking it's a nice, warm day. The snow settles on the landing board, and when the sun shines on it the bees presume that it's already summer!

To prevent this wipe the snow off the landing board, and hang sacking down the front of the hive to darken it and make it look less inviting outside.

● Mice may be in there if you have no mouse guard, tearing down the frames and eating the bees!

● Ponies being overwintered in the same field will take great pleasure in kicking over the hive.

Mid Winter Cluster

At this time of year the bees reduce themselves to some 10,000 – the older ones have died or been thrown out and there will be no replacements until the queen starts laying. In winter they will mainly stay in a tight cluster around the queen. By doing this they keep the central temperature of the cluster up to 65°F (17°C) to survive up to Christmas, and then push it up to 92°F (33°C) when she starts laying for the new season in January – traditionally the coldest month.

The bees beat their wings and 'shiver' to keep up this temperature while circulating round and round in the cluster. This takes energy, so the cluster has to move around the frames consuming the stores in a methodical pattern. If the stores run out they will cluster under the candy; otherwise they just go up to it as they need. Remember that there must on no account be a queen excluder to separate the queen from her workers. They will not leave her; they will simply stay with her and starve.

Mid Winter Flying

The bees will leave the hive on mild days, flying out to defecate and get rid of the husks from the pollen. They will not defecate in the hive (if they do they have Nosema), but will gradually swell up until they can get out on a sunny day. It is important that their honey is reduced to the correct water content, or they swell up too much.

The effect when they come out after a long hibernation is noticeable, for the bees can turn a line of washing completely yellow! Be aware of what you leave near the hive.

They will also go out and collect any pollen that is available, fresh being much better than their stored 'pickled' pollen when it comes to feeding the new babies in the first months of the year. The main winter sources are Snow Drops, Winter Anemones, Crocuses and Winter Aconite.

Mid Winter Check

If you can see the bees taking pollen into the hive in January, it means that all is well and the queen has started laying.

From this time onwards you should also see a growing pile of corpses round the front of the hive. These are the winter workers dying off as their new replacements are born; with more fat on their bodies their 3–6 month lifespan is much longer than the summer workers, but now their time is at an end.

Sugar candy

From E. H. Thorne (Beehives) catalogue. Thornes also produce their own special 'Ragus' candy in 2.5kg blocks.

Sugar Candy – Emergency Feed

1 pint water, 6 lbs. sugar, 1 tsp. cream of tartar. Boil water and remove from the heat. Add sugar and cream of tartar until dissolved. Return to the heat and stir continuously until the mixture boils. Continue to simmer until the mixture thickens (about 2 minutes). Drip a little onto a cold plate, if it sticks to your finger, boil a little longer. Candy should be soft, but not too sticky. When the right consistency is obtained stand the container in cold water, stir vigorously until the mixture is about to set. Pour immediately into suitable containers.

First Year Troubleshooting

What do you do when things go wrong? These are some of the questions that novice beekeepers might want to ask, particularly in their first year of beekeeping. Our answers to them attempt to describe reasons why things have gone wrong, and how to remedy the situation.

Q. The neighbours start complaining about my bees. What do I do?
A. In the first place it's usually best to be open and friendly with your neighbours – tell them what you are up to, get them interested if possible, and when you have enough give them some honey.

You must be reasonable and not create a 'nuisance'. More than two hives in a small, modern garden is becoming unreasonable, as is having them facing footpaths, looking through frames when there are a lot of people about, etc. In one case neighbours took an unreasonable beekeeper to court, and he was banned from keeping bees in his garden.

Q. My bees are not expanding. What is wrong?
A. They may be short of food, so feed them sugar syrup which should stimulate the queen into laying. Alternatively if the problem is a poor queen, feeding will stimulate the colony into superseding her.

Q. It's the flying season, but my colony seems to be dying. What do I do?
A. Feed the bees sugar syrup for a fortnight and see if they pick up. 90% of the time the problem is likely to lack of food. As a general tip it's worth planting pollen bearing plants around the hive. Pollen will make the queen lay early. Try crocuses, dandelions, snowdrops, winter aconite, etc.

The queen may have been killed which is often the result of over enthusiastic beekeepers pinching out queen cells, an anti-swarming policy we don't advocate. Alternatively the hive may have a problem such as brood disease or acarine mite.

Q. The bees stay on one side of the brood box. What's wrong?
A. This is most likely to happen if you put in your frames the 'warm way', parallel to the hive entrance. To cure this, take the undrawn frames out of one side and shunt the others across so that the brood becomes centralised and the whole box is eventually in use.

Q. The bees are building a lot of brace comb in the brood box. Why?
A. The spacing between the frames is almost certainly wrong. Check the metal ends are correctly positioned; make sure you are not using special wide metal ends which are designed to be used only with drawn out frames in honey supers.

Q. Why does my smoker keep going out?
A. Light it properly to start with: don't jam it too full; give it the occasional puff to keep it going; and top it up with dried grass as necessary which should keep it going all day. Your problem may be partly due to using a flame retardant cardboard.

Q. What do I do if a bee gets inside my suit or veil?
A. Don't panic. If the hive is open, cover it with the manipulation cloth. Walk away, and go indoors if possible. Remove the clothing carefully, if necessary squashing the bee between forefinger and thumb.

In this situation never attempt to take suit or veil off when standing by the hive. It's better to be stung by one bee than by 500!

Q. What do I do if a bee gets stuck in my hair?
A. Carefully comb it out. Remember that bees don't like the smell of hairspray.

Q. I've put on a brood super, but the bees don't seem to want to go up into it. Why?
A. Have you put it on too early? If not, a quarter gallon feeder of sugar syrup above the brood super will soon tempt them up.

Q. My bees won't go up into the honey super that I've just put on. Why?
A. They don't like going through the queen excluder. You must put a couple of drawn out frames from the brood super into the honey super to attract them.

Q. My bees have not covered any of the side frames, but have built upwards through the supers. Why?
A. Wait until bees cover the end frames and the wax is drawn out before putting on a super. If you super too early, the bees will go up which is their natural inclination. This is not a particular problem with honey super frames which can be moved around as they are used; however brood frames cannot be separated.

Q. When I open the hive the bees come up to get me. What do I do?
A. Always have your smoker going when working on a hive, so you can use a little smoke to subdue them. Take care not to bang the hive roof or other parts of the hive, and if the bees appear agitated stand back and leave them time to settle down.

Remember that bees become touchy at certain times when the weather or conditions don't suit them – thundery weather, end of the honey flow, etc.

Q. Some of my bees appear to be drowning on the hive roof. Why?
A. Their wings tend to get stuck down when a hive roof is wet, and then they can't move. To give them some grip paint the hive with a non-slip paint, or with clear varnish and a sprinkling of sand.

Q. Robber bees are attacking my hive. What shall I do?
A. Close down the entrance to one bee space if necessary. If the robbing is very bad, prop a sheet of clear glass in front of the hive. The robbers will tend to fly straight into it. While the resident bees find their way round the sides. Block one of these sides if necessary to make their defence even easier. Alternatively pretend that it is raining by turning on a garden spray which will give the hive an intermittent shower; your bees will be confined to barracks, but you won't get robbing in the 'rain'.

Ask yourself why the robbers have arrived on the scene. 90% of the time it is the beekeeper's fault – spilling sugar syrup, dripping honey, dropping pieces of comb, etc.

Q. My bees follow me into the house. How can I stop them?
A. Bees will follow and locate the source if they smell sugar syrup or honey, the residue of which may be left on frames and other gear. Take great care to avoid dripping, and always extract honey in a bee-proof room.

Q. The bees have swarmed in my first year of beekeeping. Why?
A. The most obvious cause of swarming is lack of laying space.

If you have bought a nuc, it's possible you took too long hiving it with the result the bees became overcrowded in the nuc box, built up queen cells, and swarmed.

MORE TROUBLESHOOTING
FOR BEEKEEPERS ON PAGE
141, TOGETHER WITH A
SUMMARY OF THE FIRST
THREE YEARS AND BEYOND.

The Second Year: Spring Awakening

In the second year your queen should be at her peak, starting laying in January and possibly carrying on as late as October. In the early part of the year you must make sure the hive has adequate stores, check its progress by watching the bees, and then give it a thorough spring cleaning on the first suitable day. By careful study of the frames you should have some idea of what kind of year to expect – hopefully one with lots of honey when your bees won't swarm.

The honeybees' favourite spring awakening must be oil seed rape which in a good year can flower as early as April. It is easy picking for the bees, producing a lot of honey which can be mixed with the main flow honeys later in the season for a mix of flavours. Changes in EEC policy means we are likely to see less rape being planted in Europe in the 1990s.

First Appearances/Moving Hives

In the early part of the year you can check the progress of the hive by watching the bees coming in and out, and seeing how fast they consume any candy. Never lift the crown board until the weather is suitably mild for you to commence spring cleaning.

Your bees should be flying out and returning with pollen on any nice, sunny days, though if it is very cold they won't go far. If your bees aren't flying and others in the neighbourhood are, put a note on top saying 'No Pollen Seen', leave candy on, and make your spring cleaning inspection at the first possible opportunity.

Having been fed down in the autumn, the bees should have had their first block of candy around Christmas. The candy gives them the necessary energy to 'shiver' and keep the brood warm in cold weather. Keep giving it to them as they consume it – you can't overfeed them in this manner and the candy acts as a kind of barometer to the state of the hive and the state of the world outside. As far as the bees are concerned it is similar to capped honey, but they still perfer the real thing.

You can buy candy in blocks from your local bee shop.

MOVING HIVES

If for some reason you need to move your hive, winter is the best time since the bees are effectively at their most dormant.

Moving house

If you are moving house, moving your hive (or hives) is fairly straight forward.

1 Close the entrance to the hive in the evening when they have stopped flying, using the entrance block or better still a strip of foam which is more likely to stay in place. Put a strap around the hive to hold it together; alternatively with a National you can use heavy duty staples to hold the boxes together, but they must be angled to prevent any movement.

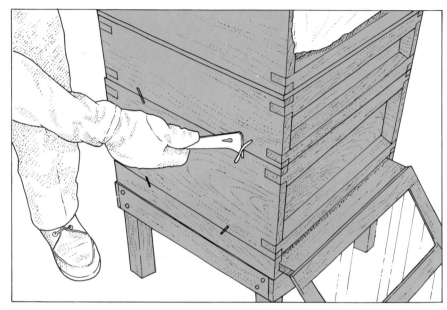

Above: Hive staples can be removed using the ubiquitous hive tool. The angle of the staples prevents movement.

Below: Single wall hives such as the National can be carried with patent 'hive carriers' for easier handling.

2 Pick it up and take it to the car (or whatever form of transport). The bees can stay in a closed hive for up to an hour, but any longer and they must have ventilation. Get a ventilation screen on the top of the hive, and cover it with newspaper to keep out the light.

3 Wedge the hive carefully in the car. It's probably safest in the boot. The frames must be aligned lengthways with the car, to help prevent them banging to and fore which could crush the all important queen.

It may feel stupid driving along in a bee suit, but what if you were involved in a crash and the hive went flying?

4 When you arrive put the hive immediately on its new site. Ideally the ground should have been prepared before; otherwise you can prepare it after the hive is established.

5 Leave the hive for 20 minutes or so, then come back and work from behind. Remove the entrance block, having remembered to place a screen in front – a piece of branch or a clump of grass – so the bees don't fly off without thinking.

What about the Rule of 3's? If you are moving less than three miles you may have problem with the bees flying back to their old site. The answer is to move them in stages, and for this the summer is best. First take them to a site which is more than three miles away. Your local association or a beekeeping friend should be able to board them.

Leave them there for six weeks. During that time all the flying bees who remembered your garden will have come to the end of their brief lives. You can then move them to your new garden, and none will realise that their hive was originally based just down the road.

Moving in the garden

The Rule of 3's – more than three miles or less than three feet – makes it clear that moving hives within your garden should be avoided if possible. However if you must move a hive the best time is the end of the winter around March.

If you move it less than three feet, you can move it a little at a time, ensuring the new entrance starts where the old one ended. If you move it more than three feet, the old flying bees from the hive will fly back to the original site, find no hive, and will die. However there should be plenty of new bees to replace them, and they will make their first flight from the new site and return to it.

If the weather suddenly turns cold soon after you have moved the hive, you may have a problem. The old bees will have flown off to the original site and died; there are not enough workers left to effectively blanket the brood and keep it warm in cold weather; and this leads to a condition known as 'chilled brood'. Many of the eggs will not hatch, and when spring returns the hive may be excessively depleted.

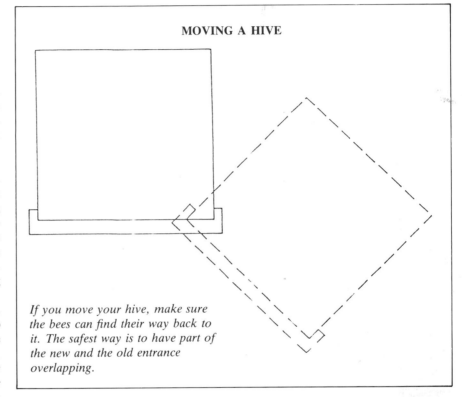

MOVING A HIVE

If you move your hive, make sure the bees can find their way back to it. The safest way is to have part of the new and the old entrance overlapping.

Four hands are needed when carrying a hive full of bees.

Spring Cleaning

Commercial beekeepers may not get round to spring cleaning every hive each year – with the early honey flow for oil seed rape they just don't get time. However for the amateur with a few hives, spring cleaning must be performed annually.

You will need some more equipment:
- Spare floor.
- Spare brood box.
- Two new brood frames, each with coloured plastic ends for the year on one side and silver metal ends on the other.
- Spare super.
- Queen paint and cage.

You will also need your glass quilt; an empty super to place the brood super on if you're spring cleaning a brood and a half (or a National roof/WBC lift will do); soft cloth; smoker; bee brush; and a pen to take notes on the condition of the hive.

When?

You should aim to spring clean on the first day when you feel comfortable in shirt sleeves during late March/April. The time to spring clean is around midday when the flying bees are out.

Make up your mind what you are going to do and don't deviate. In this case we are going to transfer the colony to the new brood box, note and record what is on the frames, and if possible find the queen and mark her. It's not the end of the world if you don't find her – if you can see pollen going in and new laid eggs in the cells, she must be somewhere amongst that mass of bees. However if you can identify her do so, for it may be a great help at a later date in the same season or during the following year.

Marking the queen

The queen can be recognised by her longer body, shorter wings, and long brown legs. Look for her in a spiral, starting on the outside of the frame and working inwards.

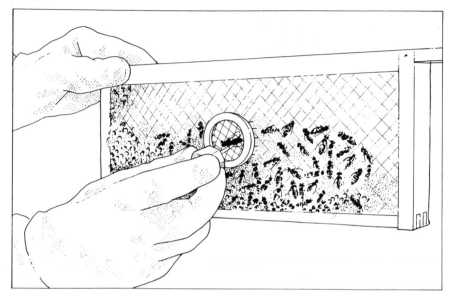

Above: The queen is isolated with a queen cage and marked with a small blob of paint before being freed.

Below: The state of the brood frames at spring cleaning helps indicate how the colony will expand and perform.

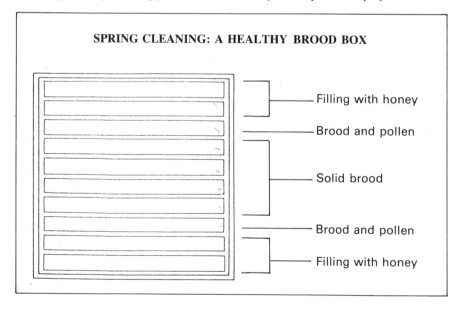

SPRING CLEANING: A HEALTHY BROOD BOX

- Filling with honey
- Brood and pollen
- Solid brood
- Brood and pollen
- Filling with honey

When you have found her, isolate her by pushing the other bees to one side with your index finger. Then cover her with the queen cage, and use the queen paint to mark her thorax with a very small blob of last year's colour to record the year she was born in.

Predicting performance

The contents of your frames will indicate what sort of state the colony is in, and what sort of honey gathering season you can expect if you continue to manage it well and the weather is benign.

A healthy colony at this stage of the season should have between two and six frames of total brood area, with pollen surrounding the brood. What you see will depend on the local flora and general weather conditions, and you should note the level of stores. If they appear low, you will need to feed the bees if the weather is too poor for them to go out foraging. If the bees are able to forage, don't feed them or they will get lazy.

A mediocre colony might have about two frames total brood area.

A poor colony might only have a small amount of brood on one frame. Either the queen is past it, or the workers are not getting enough to eat and you must stimulate them with a spring ¼ gallon (1.13 litre) feed of warm syrup made in a 1.5lb (0.68 kg) sugar: 1pt (0.56 litre) of water mixture.

The queen lays according to how much she is fed. With the influx of sugar syrup the bees will think the honey flow is on, and will consequently feed her so that she produces the bees which the colony needs.

Above: Spring cleaning being correctly done away from the hive area, with notes taken on the state of the hive. Below: If the bees are really weak and need immediate stimulation, try shaking sugar syrup over the frames.

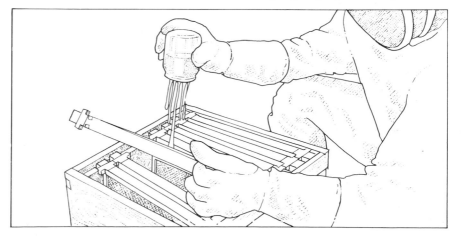

How to Spring Clean

1 Get fully dressed. Spread out your white sheet in a position which is well protected from any wind, and carry the hive over to it. Place it down with the entrance in a different direction.

Leave a skep or cardboard box at the hive site to act as temporary home for the comparatively few flying bees at this time of the year.

2 If you are on brood and a half you can firstly smoke the bees down into the brood box, to see if there is brood in the super. Alternatively start on the brood box first.

3 Lever off the super with the bee escapes in place on the crown board. Place it diagonally on an empty super (or on the roof or a lift) so the bees hanging off the bottom of the frames don't get crushed. At the same time cover the frames of the brood box with your soft cloth. You are now ready to transfer all the brood frames into the new brood box.

4 Place the new brood box and floor alongside the brood box with the colony. Cover it with the manipulation cloth.

5 Allow 15 minutes for flying bees to leave the brood box. Have your smoker lit, but as yet don't use it. Stand your two new brood frames by the side of the new brood box.

6 The aim is to transfer the frames one by one to the new brood box, keeping them in the same order. Stand behind the old box, and roll the soft cloth back to expose the first frame. Use very little smoke, preferably none, and lever both ends of the frame free. Take it out and examine it. There are unlikely to be any bees at the sides of the hive, but note down if the wax is drawn out and if there are stores on it.

7 Place the frame in the new brood box, and cover it over with one side of the manipulation cloth.

8 Work through the frames one by one. On the first and second frames you should see stores and quite a number of bees; if the queen is laying from the middle as is usual, there won't be eggs this far out.

When you have finished examining these two frames, hold them over the new brood box and use a bee brush or feather to brush the bees down into their new home. Then place these frames to one side; from now on they will be spares, particularly useful for catching swarms. Taking them out is the start of the rotational system of constantly replacing old frames in the brood box.

9 The third frame in should have stores, pollen and small numbers of eggs or sealed brood. You may even find the queen. She will be surrounded

Workers on the brood area of an old frame from a brood super. The bottom bar can be scraped clean.

and almost covered by a cluster of worker bees. A gentle blast of your breath will push the workers to one side – it's something the bees dislike intensely, so don't overdo it. Put this frame into the new brood box, up against the far side.

10 Continue to work gently through the frames, examining and transferring as you go, remembering to keep them in the same order and the same way round. Record all signs of eggs, sealed brood and pollen.

11 When you get through all the frames, fill the remaining two spaces with the two new frames with their coloured ends. From now on you will remove and replace two brood box frames each season, slowly pushing the frames from one side to the other. This technique is not necessary with super frames.

12 Take the new brood box back to the hive site. If you know the queen is in the brood box with plenty of space for laying, put the super back on top with the queen excluder between. You want to force her to lay in the brood box frames; you don't want her to lay in the super before it is necessary.

13 Some bees will be left on the old floor, and round the inside of the old brood box. Ensure the queen is not amongst them. If she is, pick her up carefully and drop her into the brood box. Place the old floor and brood box against the alighting board, and the remaining bees will gradually go up into the hive with a little help from your smoker.

14 A week after your spring cleaning, take a quick look at the middle frames in the super. If the queen had laid, the nurse bees will stay until the baby bees are hatched which is a 21 day process. Wait until the brood is hatched; then take off the super, let the bees find their way into the hive, and replace it with a new super and frames.

If there is no brood, remove the super and place the glass quilt over the brood box so you can keep an eye on the colony's progress. Leave the super by the hive entrance so the bees can find their way up into the brood box.

Above: Remove brace comb with the scraper end of your hive tool, and store it in a sealed container (wax moth proof) for future use as beeswax as described on page 160.

Below: Brood frames should be examined one at a time, with two being replaced. Note the manipulation cloth covering the frames, and the smoker ready for use if required.

It is easy to miss the queen when you are inexperienced. However if you are spring cleaning a brood and a half she may be in the super, and if you did not find her in the brood box you should check to see if she is there.

Examine the frames one by one, transferring them to the new super in the same way as with the brood boxes. The only difference is that you don't change two frames for new frames.

Once again look for brood, eggs and stores. If you find the queen, place the cage over her and leave it there. You want to get her back into the brood box where she should be laying.

Transferring the queen

1 Cover the brood box with the soft cloth – nothing else.

2 Place the old super on top with the queen frame still in it and the queen imprisoned by the cage.

3 Lift one corner of the soft cloth to expose the brood box frames. Take off the queen cage, brush the queen down, and drop the soft cloth back over her to force her down. Remove the old super.

4 Place the queen excluder over the brood box, and then put on the super with the old frames plus the one missing frame. Ensure that you scrape the top bars of both brood box and brood super clear of excess build-up of wax or propolis.

Cleaning the gear

The old brood box, floor and super must be cleaned immediately. Scrape off all the easily removable debris. Some of this should be bottled, and sent to the local association who will normally be able to test for varroa. Scrape and flame all the surfaces until all debris and disease is removed.

You can store your two old drawn out brood frames in the old brood box. They will come in useful for holding a swarm if you catch one, or a nucleus if you decide to make one. A queen will lay straightaway in drawn out frames.

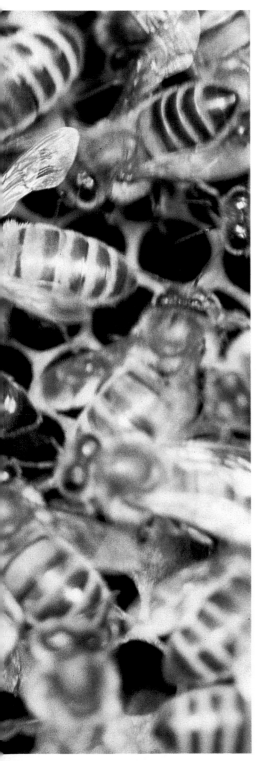

Below: Can you spot the queen? She is easily recognised with the beekeepers blob of colouring.

Acarine mite

Acarine mites debilitate the bees, and can spread rapidly when the bees cluster in cold weather. When you are watching the bees early in the season, you may see bees with a split 'K wing' finding it difficult to get into the hive and climbing up blades of grass. This may be caused by acarine mites.

If you see one or two bees in this condition, note it down. If you see a lot, get them analysed via your local association or shop, and then treat them while you spring clean as soon as you can safely open the hive.

Treatment

Treatment should be according to the instructions issued with whichever chemical is used to kill the mite.

The use of a Folbex VA strip is a popular method. The basics of using one are to take off the crown board in the evening; put an empty super on top of the hive with the crown board on top and a Folbex VA strip pinned underneath; block the entrance with newspaper (the bees will eat their way out if you forget to unblock them); and set light to the strip which will smoulder until its fumes have killed the mites.

USING FOLBEX VA

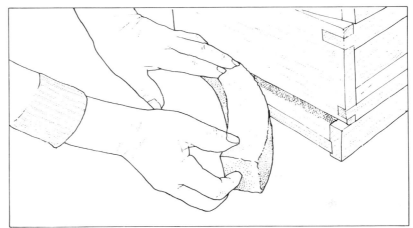

When using Folbex VA or similar smoke treatment you can close off the entrance with foam rubber or newspaper.

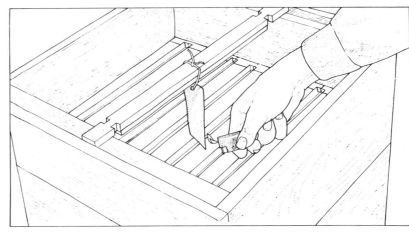

The Folbex strip is suspended in an empty super over the brood frames, and lit with the roof replaced.

Getting a Second Hive

Your second year of beekeeping is a good time to get a second hive, and there are two good reasons why you should do so:

1 If you have one hive it is always possible that you will lose your colony – a hard winter, disease and swarming all take their toll. If you have two hives, one or both will normally survive. If both are weak after a hard winter, they can be amalgamated.

2 Two hives are not much more work than one – say another 25%. However two could be twice as productive, giving a normal family excess honey rather than just enough.

You are halfway to a second hive with the new floor and brood box you needed for spring cleaning. The extras to make a full hive are a roof, a glass quilt, new frames for the brood box, and extra lifts if it is a WBC. You will also need more supers and the frames to go in them, but probably not until the third year. It goes without saying that you should stick with the kind of hive you already have.

Siting a second hive

You can see our recommended way of siting a second hive on page 51. The points to remember are that you want the two hives adjacent to one another so you can work them at the same time, and their entrances should be angled at 45–90° to one another.

There are three ways of getting your second colony, which are all covered in detail on different pages.

1 Buy a nucleus as for your first hive (page 54).
2 Catch a swarm (page 130).
3 Make up your own nucleus (page 120).

Buying them is of course the soft option, and we would recommend you try the other two more interesting methods!

Above: Side by side – National floor, brood box and roof; and floor, twin brood boxes and roof which is here being used for queen rearing.

Right: The more hives you have, the more forage you need. Plenty of suitable trees, shrubs and flowers are needed to keep the bees happy.

The Second Year: Lots of Honey

The second year of your colony should be the best for honey collection. If you live in an area where they plant oil seed rape you can expect a lot of honey early in the season. Later in the summer you can then harvest the main crop, or alternatively decide to make up a nucleus to establish a second colony. Planning ahead to avoid the effects of swarming in your third year also becomes very important.

Not perfect but not bad, this frame taken from a honey super is worth a few pots of honey. It is almost 100 per cent capped over with wax, the bees' means of storing it for harder times ahead. This means it is fully ripened with all excess water removed to prevent fermentation. In this state it will keep perfectly for years. The wax cappings are also useful; store them as future beeswax for candle making and other uses.

The Early Flow

The early honey flow in the UK is at present dominated by oil seed rape, and will continue to be while the farmers plant it. If you do not live in a rape area, this section does not apply but should be understood nonetheless.

Oil seed rape is supersaturated with sugar, and it gives the bees better return for their foraging than any other crop. This is bad luck for fruit trees which have nectar that is very low in sugar content. The rape flowers come first, and once they are working it bees which are 'plant constant' will not abandon it.

What happens when farmers want their orchards pollinated? The only way is to take a hive (or hives) and put it amongst the fruit trees just as they flower. Bees by nature will go to the nearest blossom that is available, and will not look any further. If they have been working rape, they must of course be well out of range of their previous working area or they will return to the rape, and then return to the original hive site where no hive will await them.

The main rape crops are the autumn planting which generally starts to flower in May and has finished by the first week in June; and the spring planting which generally flowers from mid to late June. The nectar secretion of the rape flowers which triggers this helter-skelter honey flow will depend on the weather; if it is unseasonably cold nectar will not rise.

You should have your spring cleaning done before the oil seed rape honey flow commences. Sometimes this is not possible, but you should at least change the floor and then do a proper spring cleaning if there is a convenient break in the honey flow or in the autumn.

Unlike normal honey, rape honey can become rock solid within 10 days of being stored in the hive, and you need a plan of campaign to ensure you successfully remove it.

Other early season honey

The honey flow from other sources pales into insignificance beside oil seed rape. If your bees don't have rape available, you leave the small amount of honey they have gathered for their own stores and wait until the main summer flow to take some for yourself.

In your second year you should expect at least 30lb (13.6kg) of rape honey (one full super) from one hive. When you have more drawn out frames the bees can concentrate on honey storage rather than wax building, and you could usually expect to take 30–90lb (13.6–40.8kg) from a hive depending on the variables that are a part of beekeeping life. The record we know of is 212lb (96.16kg) of oil seed rape honey from one hive!

A trailer taken to a rape field to make the most of the early flow. Note the bee entrances cut into the sides.

Though not a classic honey, it is a pleasant honey to eat. It does however become more and more solid with time, and in its time has bent many a spoon. You can soften it by standing the jar or tin in warm water. Ensure the temperature does not exceed 145F (63C) over 30 minutes, beyond which the enzymes of the honey will start to break down, losing its health giving qualities.

Right: Unseasonably hot weather during the early season brings bees out onto the front of the hive for some fresh air – it can also lead to some unseasonable swarming. The quarter gallon sugar syrup feeder is being used to feed a weak nuc in the background, which despite the weather has not prospered.

Below: Beehives are traditionally taken to orchards for their valuable pollination at blossom time. The nectar has too little sugar to compete with oil seed rape as the major early honey flow provider.

Oil Seed Rape Management

The oil seed rape honey will need to be stored in a honey super placed on top of the hive, but separated from the brood by the queen excluder. With drawn out frames from your first season the bees can fill a super in four to ten days once their rape honey supply comes on tap, and as we have said this honey must be removed promptly if you wish to spin it off. In a non rape area there is less problem, less honey. You simply stack the hive up with two or three supers, leaving them until the end of the season when all the honey can be removed in one go. Similary rape honey can be left if you're going to cut it out of unwired 'cut comb' foundation.

Keep an eye on the bees' progress by watching through the glass quilt. If the bees start building 'brace comb', out under the glass quilt it means they have run out of honey storage space in the frames and are building up extra wax to accommodate it – time for you to remove the super and the honey.

During these few critical weeks you should make yourself available for a spinning-off session each weekend, being ready to take off and spin-off the super at the point when it is thick enough not to drip from the frames.

Technique
In our spring cleaning we have presumed the hive has stayed or reverted to a single brood box, plus queen excluder and honey super on top. When the honey super is due for removal, proceed as follows:

1 Get a spare super ready with drawn or undrawn frames and a clearing board fitted with Porter bee escapes on top.
2 Go down to the hive in the evening. Plug the glass quilt with the bee escape and gently remove the honey super, laying it diagonally across the upturned roof or a lift. As you do this, quickly drop a soft cloth over the queen excluder to keep the bees in the brood box down. While doing this it is safest

to work over a sheet of plastic to avoid dripping honey on the ground.
3 Put the new super on the queen excluder, pulling out the soft cloth. Place the honey super on top of that, and then reassemble the hive.
4 By next evening the bees in the honey super should have gone down through the clearing board, so you can then take it into the house to spin off whatever honey is available. If they don't clear down, the colony is probably overcrowded and will need more space – ie another super.
5 Leave the hive for a few days to settle, then go back down. Remove the new super in the usual way, placing diagonally on upturned roof or a lift. Quickly cover the queen excluder with your soft cloth; prise it free with your hive tool; pull it out and leave it placed against the alighting board at the front of the hive for any bees that are on it to run up.
6 Place the new super on the brood box (removing the soft cloth); the queen excluder on next; and the super

Rape must be spun off as soon as it is capped over; once crystalised it can only be cut out or scraped out.

with the spun-off honey on top of that.

Reassemble the hive with glass quilt, bee escape removed, roof, etc. You now have a brood and a half, plus honey super separated by the queen excluder.

If the colony was already on a brood and a half, you cannot remove the brood super which will contain a mixture of rape honey and brood. If you dare take two or three side frames out, shake off the bees and claim the honey, but be very careful to check that you are not also spinning off brood. This is a common mistake, which of course effectively destroys some of your future main flow work force

Super full again
You are unlikely to get another super full of rape honey, but if you feel it needs to be spun off take it indoors following the normal honey super removal procedure, and check through

the frames. Spin off any that are ready, and put the super back on the hive with a mixture of the wet frames and those which were not sufficiently ready.

The bees can then continue working on it, using it to store their main flow honey later in the season. Once the rape flow is finished, you can just keep adding supers to the hive – indeed apart from the problem of rape crystalising so rapidly it is best not to remove honey until the season is over since every time you open the hive it disturbs and sets back the bees.

No time for regular inspection?
What if you are unable to work on the hive each weekend during this all important early season honey flow? The bees won't mind, but you will have the problem of returning to find a super packed with honey (hopefully) that has set too hard to be spun off.

As we have made clear, if you want to spin off your oil seed rape honey, you must keep taking it off at regular intervals. If you have to go away for a month you may leave before the flow starts and return when it is all over. To ensure you get a good honey crop in spite of your absence, we would recommend leaving the hive as: Brood and a half; queen excluder; honey super with thin unwired foundation; 2nd honey super; 3rd honey super.

When the honey goes solid you can't spin it off, but you can cut it out of unwired wax and get cut comb. Whether you put wired or unwired wax in the 2nd and 3rd honey super depends on how long you will be away and how important you rate spun-off honey

Have the bees gone mad? They build some magnificent edifices!

which does after all preserve your drawn out frames. Allowing four to 10 days to fill a super, you have to guesstimate when the honey is likely to set hard.

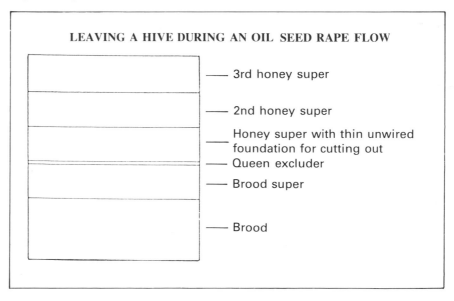

LEAVING A HIVE DURING AN OIL SEED RAPE FLOW

—— 3rd honey super

—— 2nd honey super

Honey super with thin unwired foundation for cutting out
—— Queen excluder
—— Brood super

—— Brood

The Main Flow / the June Gap

When the spring honey flow ceases, there may be a break with no nectar producing flowers until the main honey flow is triggered by nectar becoming available once again. This is traditionally called 'the June gap'. It can also be caused by unseasonably cold weather when the nectar fails to rise.

In an area with no oil seed rape it is unlikely to be a problem. The spring honey flow is comparatively weak since it comes from secondary sources with a high water content. You leave the honey untouched so the bees have plenty of stores to tide them over, and wait until the main flow before you take off honey.

In a rape area with both autumn and spring sown rape (flowering in the spring first, and then in late spring/early summer) there is unlikely to be any gap at all. However if the farmers only sow autumn rape, you may well have a very strong hive that suddenly finds its nectar and pollen sources cut off. You have removed their rape honey to prevent it going solid; they have no food coming into the hive and could starve, and it is at this time that they think about swarming.

The answer is to feed them. If there are no flowers out, try them with a quarter gallon feed of sugar syrup in a weak 1lb:1pt (0.45kg:0.56 litre) ratio. Put it on late in the evening, don't drip, and close the entrance down – no nectar available means this is a high time for robbing.

The bees won't touch this weak mixture if they don't need it. If they do and it goes down rapidly it means they are hungry. Keep feeding them until nature turns on the nectar sources once again.

THE MAIN FLOW

Depending on where you live, the main UK honey flow runs from the end of June onwards. There will of course be regional variations between south and north, though as always dates will

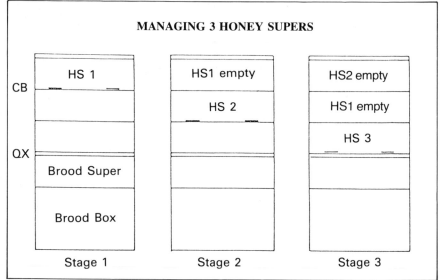

MANAGING 3 HONEY SUPERS

Stage 1: *The clearing board (CB) is inserted, and the top honey super (HS1) is removed for spinning off some 24 hours later when the bees are cleared down.*

Stage 2: *The clearing board is inserted under the second honey super (HS2). This is then removed after another 24 hours. HS1 can be put back on top of the hive before this in its 'wet', state to be cleaned up by the bees. It can be temporarily removed to take off HS2; or more easily you can combine the two jobs in one, putting back the empty HS1 as you take off the full HS2.*

Stage 3: *The clearing board is moved to its third position, over the queen excluder (QX) on top of the brood. Once the third honey super (HS3) is removed for spinning off, the bee escapes are taken out of the clearing board to let the bees get to work cleaning up the wet supers and taking remaining honey down to the brood as quickly as possible. The empty HS3 is eventually put back on the top, so you end up with HS3 on top of HS2 on top of HS1. These can finally be removed when you winter down the hive.*

depend on the weather.

The most important crop used to be clover which is unfortunately not planted much by farmers, though beekeeping enthusiasts with sufficient land plant their own American Sweet Clover which makes a lovely honey. Instead you should look out for the flowers appearing on primary sources such as blackberry bushes, lime trees, dandelions, etc – a full list of major honey plants and their flowering periods is given on pages 75 and 76, and this splendid mixture of natural sources is what makes British main crop honey so

good, despite the claims of other countries!

The main harvest

The end of the main honey flow generally comes when the blackberries cease flowering, after which the bees should be allowed to keep whatever they find to forage. You should reckon on taking all the honey off by the end of August, with a couple of weeks leeway depending on where you live and how the weather has been behaving.

The only exception is heather which runs very late on the Scottish and Welsh uplands. It only secretes nectar at a high altitude, and special techniques and bees which are specially built up for the purpose are needed to work it in commercial applications.

When you take off honey at the end of the season the bees know it is robbery. Their nectar sources are rapidly dwindling, and they are likely to become very bad tempered. Always cover yourself up well.

Go down to the hive late and put on the clearing board. Twenty-four hours later lever off the honey super, taking great care not to drip honey when the super breaks free. It is easier to do this with an assistant, and the surrounding area should be covered with plastic sheeting to catch any wayward drips. You may need smoke to keep bees away, but take it easy or you will get smoke flavoured honey!

If you have more than one honey super piled up, you can take off the whole lot in one go. However the bees get less hot and bothered if you remove them one at a time; it's more effort but is better practice.

A useful tip is to very gently block the hive entrance with net curtaining before levering off the super. That way you won't risk a horde of bees following you into the house with your bounty. Put the super in a bee tight room; wash your hands to get rid of any honey smell; and then return to the hive to put on the roof and unblock the entrance.

The hive entrance can be temporarily blocked with net curtaining, but be sure to remove it as soon as possible.

Putting back supers/Wintering down

When you've spun off the honey, put back the wet super with the bee escapes removed from the clearing board. If you are dealing with two or more supers, put them back in the order shown in the illustration on page 110.

The entrance must be closed to no more than a few bee spaces to prevent robbing. As at the end of your first year, leave the wet supers on until the bees have cleaned them out, and then remove and moth proof them for the winter. They can be left on the hive until you winter it down if this is easier for you.

In the second year you should not

This hive has a brood and a half, and four supers. If the bees were strong enough you could theoretically build the hive up to the sky. In practice this is as much as any colony is ever likely to fill up with honey, and watch out that the wind doesn't blow the whole lot of it over!

need to take honey from the brood super. Give the bees their winter feed, and when it is stored check the brood super. If it is more or less full leave them on a brood and a half for the winter; if it is fairly empty remove it, and leave the colony to winter in the brood box.

Problems with a Poor Harvest

You may get anything from 5 to 100lbs (2.2 to 45.3kg) of honey from a single hive. In a rape area 60lbs (27.21kg) would be considered average – 30lbs (13.6kg) from the spring rape and 30lbs (13.6kg) from the main flow.

A bad year?

If your honey harvest is below average, you should first check with other beekeepers in the area. It could simply be that it is a poor year, for beekeeping years are generally cyclical – very good/poor/mediocre/very good/etc.

A bad area?

An even simpler answer may be that your hives are in an area which is no good for bees, a fact which would not necessarily become apparent until your second year.

It is a general misconception that the best place for a beehive is in the country. London produced the most honey per hive during the war, and apart from the absence of rape may well continue to do so with so many small gardens and parks available for foraging. In the country intensive farming has ripped out many of the hedgerows, leaving miles of open fields and nothing for the bees. If you decide that your bees are in an area with insufficient forage, the only answer is to move them.

Bugs in the hive?

The hive may be affected by disease. Acarine mite may have got in during an unseasonal cold, wet spell; feeding too much sugar syrup early in the season may have helped give your bees dysentery or Nosema; wax moth may have colonised the frames; your bees may even have foul brood, though this is unlikely.

Prevention and cure of the various pests and problems can be found in the section on page 170.

Bees usually do well in an urban environment. This mini apiary is protected by fences and trellises, and the owners are real experts who know how to keep their colonies under control. For the novice one or two hives in this space would suffice.

Spraying

DEATH BY SPRAYING

Food & Environmental Act 1985: 'Anyone selling, supplying or using pesticides is now under general obligation to take precautions to protect the health of human beings, creatures and plants, to safeguard the environment and avoid water pollution'.

It is possible that your bees have been sprayed to death on the fields while you know nothing about it. To make sure that this cannot happen you must communicate with any farmers within bee distance and request that they notify you of all intended noxious spraying.

The bad old days of thoughtless spraying do appear to be over, helped on their way by a lot of publicity and several court cases. In one, Sussex beekeepers were awarded £10,000 damages against a farmer after their colonies were wiped out as a result of Hostathion spraying. Farmers are now usually 'bee wise', and fully aware if the substances they are spraying are dangerous.

Farmers should be encouraged to confine their spraying to early morning and late evening. If the farmer intends to spray something unpleasant early, you should block the hive entrance with netting the night before so the bees cannot get out – remember that they go to work about 5am in the high summer! Once the spray has settled it is not dangerous, so you can safely unblock the hive an hour after the spraying has ceased.

What do to

The BBKA has a 'Spray Committee' as do most local associations. The dangers are greatest in an oil seed rape area since this is a crop that can be infested by pests. The pests have to be sprayed, and the bees who are certainly not pests are in danger of getting sprayed as well. It is a relatively new problem for farmers, since bees are not attracted to their more traditional crops. However

with EEC policy constantly changing there may well be a tail-off in rape planting in the near future which will go some way to remove any problems.

Hostathion is the most lethal insecticide that has been used so far. It is designed to be sprayed at 100% petal fall when there is no danger to bees. If used incorrectly – on open blossom (and worse still at midday in the sun) – it will kill every insect it comes into contact with and almost certainly wipe out your bees.

At the time of writing Hostathion was rapidly being superseded by bee-safe sprays, and the National Farmers Union had adopted a voluntary scheme to notify beekeepers if harmful substances were to be used. The great majority of farmers are likely to stick by this, but if you feel your local farmer is uncooperative and endangering your bees, you should contact the local association Spray Committee. It is obviously most productive to strike up a friendly relationship with him – after all his flowering crops should benefit from your bees.

If you suspect something is wrong, watch the bees going into your hive to check for possible damage towards the end of flowering. If you then come to the conclusion they are spray damaged you may be able to claim damages. You should inform your Spray Committee immediately, inform the farmer, and contact the Ministry of Agriculture, Farming and Fisheries.

You should then record the event.

Make a note of prevailing wind direction and strength. Sprayed bees flying against the wind will almost certainly have died in the fields. If there are dead bees around the hive however, photograph them with a copy of a newspaper to show the date. Don't wait to do this since the birds will soon come to eat them, insecticide and all! If your hives are in a field, you should also try to identify the area with a witness included in the photo.

Take samples and send them recorded delivery to National Beekeeping Unit, Luddington, Stratford-upon-Avon, Warwickshire GV37 9SJ for analysis (there may be a charge). You should aim to send around 200 dead bees which will fill four match boxes – don't put them in airtight containers or they will decompose too rapidly. You can also seal more dead bees in a plastic bag and put them in your freezer which will help to retain the traces of poison in their bodies.

Finally check if other beekeepers in the area have suffered losses. If the farmer is quick to offer you compensation, accept nothing until you have checked the correct BBKA scale of settlements.

Your bees could also be affected by gardening enthusiasts spraying their plants and fruit trees. Good neighbourly relations is the best answer if this looks like becoming a problem, though with more awareness of the dangers of sprays some of the more noxious ones are no longer available.

PROOF OF WHERE AND WHEN: If your bees have apparently died as a result of irresponsible spraying, it's worth taking a photo of dead bees with a copy of the day's newspaper as shown. Temperature and climatic conditions should also be recorded. You would be wise to notify other beekeepers, and if possible trace the sprayed crop by estimating the destination of the bees' flight path. If sure of the facts you can approach the farmer.

The Second Year: Forward Planning

A queen will normally last three years, with the result that the third year is often – but not always – the swarming year. From the beekeeper's point of view this is best prevented. If the bees are left to their natural inclinations you could lose the whole hive, so the answer is to plan ahead of them and in effect ensure they never get to their third year.

At this stage we recommend you read the rest of the second year section, and then read ahead into the third year so that you understand the principles of swarming and how it relates to the second and third years. From now on beekeeping becomes more highbrow!

The fantastic things bees get up to if you leave them to their own devices! This swarm has started building a wax city in the tree. It is a beautiful sight, but no help for your honey production which by this stage of the game requires foresight and forward planning. In days gone by 'honey rustlers' would steal these combs from the bees and wring out the honey.

Keeping the Queen in Charge

At this stage you have a mature colony and should be considering the possibilities of the bees swarming from the hive; superseding their queen voluntarily; or falling prey to queen failure which will necessitate the building of 'emergency cells'.

Swarming

Your bees may swarm in the second year because of overcrowding. June or thereabouts is the likely time – if it happens, turn forward to page 126 to find out how to cope with a swarm!

Hopefully they won't swarm at this stage, but in the third year they may well do – the queen is becoming old, half the colony wants a replacement, and the hive's natural answer is to swarm and it may continue doing so until the hive is swarmed out.

If they are going to swarm, the bees will build 'swarming cells' all over the place for the new queens they will throw 'casts' with (smaller swarms following the prime swarm with the old queen.) If you are on a brood and a half, the easiest way to check for them is to put a wedge between the brood box and brood super – you will then be able to see if swarm cells are hanging down under the frames. On a brood box alone you can only see by taking out brood frames.

Supersedure

Alternatively a vigorous colony may decide to surpersede the queen, in which case they won't swarm and if you spot what's happening you can simply leave them to their own devices.

As the end of the season approaches

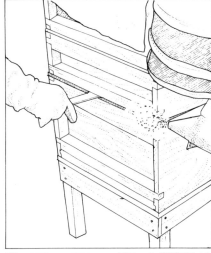

The easiest way to make an initial check for queen cells is to lever up the brood super (above) and wedge it up (below) to see under the frames.

the workers will build about three perfectly formed queen cells on the front of a frame. If you see these, leave them well alone because they will produce the best queen of all. The bees do this because the queen is getting older, and is therefore not producing enough pheromone – a drug which she secretes to keep them under control.

They will only allow the best of these young queens to live, and she will mate and lay eggs side by side with her 'mum' until the colony is satisfied with her. They will then bump off 'mum'.

Some colonies regularly supersede their queens in this manner, and never swarm. If you have such a colony you should count yourself lucky, and use it to requeen other 'swarming' hives when you get more experienced.

Emergency queen cells
Your queen could die at any time – it may even be due to your own careless handling!

If this happens alarm bells will ring in the hive. The workers will draw out the wax surrounding a young larva and start feeding it Royal jelly, usually making one or two very small queen cells. The older the larva, the less good the resulting queen will be and she may be dubbed a 'scrub' queen.

Any queens which are produced in this fashion are allowed to mate and lay to ensure the colony can survive. Meanwhile the workers build proper queen cells, and hopefully a good queen emerges and will supersede the scrub queens. However the whole colony will have been thrown out of sync by this emergency, and will be severely weakened.

The Answers
If you spot supersedure cells, the colony will rejuvenate itself with a young queen without your interference.

However you are unlikely to be so lucky, and towards the end of your

This early swarm landed in a willow tree overhanging a pond – the authors are here seen trying to catch it!

second season you would be wise to take one of two actions designed to keep your colony vigorous and prevent it swarming with the possibility of losing your bees.

• Replace the queen at the end of your second year. This is called 'requeening'. It is the best way if you want to keep your main flow honey crop and your bees, and should be used cyclically on all colonies with a queen in her second year.

• Alternatively 'make up a nucleus' by dividing the colony into two. This will give you a second colony with a young queen for which you will need a new hive; and will weaken your first colony to the extent that it is most unlikely to swarm that year. It can only be done if you are willing to forego harvesting the main flow honey.

Requeening

The main extra equipment required for requeening is a nuc box or a spare brood box with a frame shaped dividing board. Your aim is to encourage the hive workers to start making queen cells, and the best time to do this is June.

1 Uncover the brood box and take out a frame with day old eggs, a frame with sealed brood, and two frames with honey and pollen. Put them in the nuc box, and replace with new frames at one end with the correct colours for the year.

2 Pull out some of the other frames and brush the workers into the nuc box, making sure the queen stays. You must have enough young workers to look after the brood. The foragers will return to your hive ensuring the honey flow continues.

You now have the hive with flying bees and the queen; and the nuc with nurse bees caring for the two frames of brood which were removed. Note that the nuc entrance is angled away from the hive.

If you couldn't find the queen, ensure there is at least one frame containing day old eggs in each colony.

3 After three days feed the nuc with a quarter gallon of sugar syrup (1.5lb sugar: 1pt water), [0.68kg sugar: 0.56 litre water] remembering to close down the entrance to guard against robbing.

4 Leave the nuc for a fortnight, and then inspect the brood frames for queen cells. If none have appeared, introduce another frame of eggs from the hive brood box and wait again. Alternatively if any queen cells are present in the hive, a frame with them on could be introduced.

5 Let the queen cells in the nuc grow, and don't touch them. The new queen will hatch out (the bees will decide which one lives if they make more than one), hopefully get mated with drones, and take command of the nuc. She should then start laying her own eggs.

Don't expect any honey from this nuc, and keep feeding it as necessary with an eye on the nectar supply.

Sometimes the new queen is not mated quickly enough. Bad weather may prevent her getting out to meet the boys, and if she's not mated properly within 10 days she may have insufficient sperm and become a 'drone layer'. Too many drones and not enough workers spell disaster for a colony; this and other problems are dealt with on page 134.

6 At the end of the summer in August you must arrange a first and last meeting between the two queens.

If you have been using a nuc you will need to transfer the new queen and her cohorts into a brood box. Late in the evening, place it alongside the hive with the entrances facing the same way. Expose the frames, cover them with a sheet of newspaper which has been pricked several times with a pin, and place the brood box with the old queen on top. Nature dictates that the young queen will go up and kill the old queen, and the bees will amalgamate happily.

The result is that you are in effect back to year one with a young queen ruling a strong colony which should have no interest in swarming in its new queen's first year.

Buying a queen

If you don't succeed in requeening in your second year, you can wait until the spring of the third year and before the swarming season buy a young queen and amalgamate her with the hive.

This is not a good method for those who are relatively inexperienced. The technique is:

1 Kill the old queen by the simple expedient of squashing her.

2 Slip the new one down between the brood frames. To do this you use a special 'introduction cage' to hold her, or a hair curler with candy stuffed into the ends will do as well.

3 Ninety per cent of the time the resident bees will reject her, kill her, and start to make their own queens. Therefore this is not a good method!

Buying a nuc

Alternatively you could buy a nuc, as in your first year.

When you do your third year spring cleaning, put the old queen plus colony in their brood box on top of the young queen plus colony in a brood box, using the same method as we have already described in Step 6 above.

The new queen will kill the old one, and the new and old colonies will amalgamate under their young leader.

Right: Buy a nuc or make your own new queen? A nuc box which is essentially a half size brood box soon becomes a useful beekeeping accessory which can be used for making new queens, making new colonies, and hiving swarms.
Below: A new queen can be introduced between the brood frames using an introduction cage.

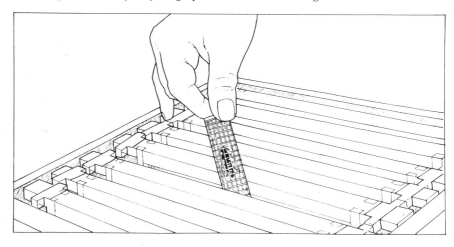

Making a Nuc

Rather than sticking to one hive and requeening in your second year, we would recommend making a nuc. This will give you a second colony with a new young queen, while the original hive is so weakened that it is most unlikely to swarm in the third year by when it should supersede or be requeened.

However you can't have both more bees and honey. If you live in an oil seed rape area you can get a good crop of rape honey in the spring; and a good crop of bees in the summer which will give you your second hive.

Technique

When you've taken off the spring rape honey in June, that's your lot! To make a nuc the existing colony must be strong with a large brood area. If there is not much brood, there is no point in depleting it further.

The existing queen should also be marked so you can recognise her. Furthermore you must ensure that drones are flying – they are needed to mate with any new queens within 10 days of their hatching.

1 Move the hive to a new site to get rid of the flying bees. Place it and the new brood or nuc box alongside one another on a white sheet, with their entrances aligned with one another.

2 Open the brood box and remove a couple of frames with honey and pollen; and a couple with brood, one of which must have day old eggs. Any queen cells on these frames will be a major advantage. Locate the queen who must stay in the hive. If you can't find her, she may be inadvertently transferred to the nuc; in that case she will start to lay in the nuc, and the hive will need day old eggs to produce queen cells and survive.

3 Put these frames in the new box together with two spare drawn out frames. If using a brood box, use a dividing board (or a frame feeder) to close off the unused space.

4 Push the remaining frames in the

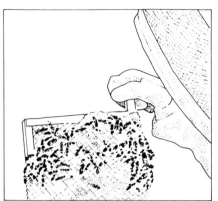

Examine the brood frames. The nuc must have one with day old eggs; better still with closed queen cells.

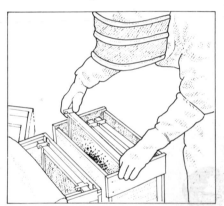

Fill the nuc with frames containing brood, honey and pollen, plus a couple of spare drawn-out frames.

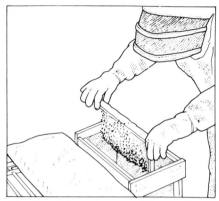

The day old egg frame will be used to make the new queen. Ensure you do not transfer the existing queen with it.

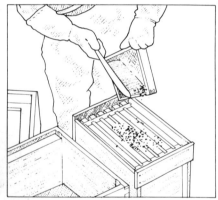

You must have sufficient bees to look after the brood in the nuc – if necessary brush them off brood frames.

original brood box together, and then introduce new frames into one side with the correct coloured ends to show you which year they were introduced. Ensure there is not brood against the sides of the box and brush in some nurse bees if necessary.

5 Put the hive back on its original site, and put the nuc on the site of your second hive. The entrances should be angled away from one another and closed down to a few bee spaces.

Give both nuc and hive a quarter gallon sugar syrup feed three days later, ensuring the entrances remain reduced to guard against robbing by other marauding bees.

Waiting for a queen

Keep feeding the nuc to help stimulate the bees into queen production. Check the frames and if there are no queen cells after 14 days, try introducing another frame with day old eggs from the hive. Eventually you should have queen cells and a queen.

Once the queen is born, if necessary continue to feed the nuc to stimulate her into laying. Keep an eye on pollen being taken in by the foragers – if there is none it means the queen is not laying and there are no larvae to feed. Hopefully the nuc will expand rapidly – if the bees start to look cramped, give them more frames.

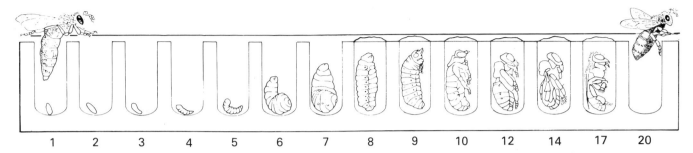

| 1 | 2 | 3 | 4 | 5 | 6 | 7 | 8 | 9 | 10 | 12 | 14 | 17 | 20 |

Once it's up and running the nuc can move into your second hive. You now have two colonies for the price on one, but because you have depleted the first colony to make the second, the chances of third year swarming are greatly reduced.

The tale of day old eggs
The queen lays some 1,500 eggs per day at the height of the season. A newly hatched egg can become a worker, drone or queen. The bees make the bees that they need – the queen to lay the eggs, drones to mate with queens and balance the hive, and workers to work the hive.

On the first day the eggs stand upright in their cells and once they become larvae are fed Royal Jelly; by the third day of feeding the nurse bees will have switched their diet to pollen and nectar if they are to become workers or drones (the queen decides whether to lay female or male eggs), and they'll be lying down in their cells, with drone cells being recognisable by their dome shaped covers.

If on the other hand any of the eggs are to become queens, the nurses will continue to feed Royal Jelly and will build a pronounced acorn shaped cover for the 'queen cell' which is big enough to allow her to develop.

The development of a worker from egg laid by the queen (day 1) to the worker emerging ready for hive duties by day 21. Note that the worker cell is capped over at the end of the first week. Drones emerge from their slightly larger cells in 24 days. Queens emerge in 15–16 days, developing in very large cells that protrude from the combs. Both workers and queens develop from fertilised eggs, but the queen is fed on royal jelly throughout her development.

An area of enlarged drone cells is clearly visible on this brood frame. You can also see sealed honey, some worker brood, and a lot of bees!

Requeening Two Hives Etc

The requeening technique is simple and ensures that a colony is always thriving, strong and very productive; rather than ailing and about to swarm under a weak queen. It accomplishes two important things.

●Firstly it weakens the hive by removing brood that makes baby bees and the nurse bees to look after them, thereby hopefully ensuring that it won't swarm. However in the meantime it is still inhabited by a full complement of flying bees who are gathering honey for you, without so many hungry little bees to consume it.

●Secondly it provides a new queen to take over at the end of the honey season and get the colony back into full honey production the following year, rather than declining and/or swarming.

Once you have sufficient colonies from making up nucs (page 120), you should requeen as regular practice with all second year colonies. Wait until the queen has produced a superb colony for honey collecting, and then replace her at the end of that season with a younger, more dynamic queen. With two colonies you should be able to requeen alternate colonies on alternate years, using a simple rotational system.

Requeening two hives

You may have two hives running in concurrent years, and need to requeen them both in the same year.

To do this take day old egg frames from each hive and make up two nucleii, either using two nuc boxes or a brood box with a dividing board between the two sets of frames. The latter can be confusing for the bees with the entrance to the two nucs on the same side, and one of the new queens may go back in the wrong side after her mating flight. Specially designed brood box floors with entrances at back and front are available to avoid this; or move the frames from one side of the dividing board into a separate box as

```
┌──────────┬───┬──────────┐
│ Entrance │   │          │
│          │ D │          │
│          │ i │          │
│          │ v │          │
│          │ i │          │
│          │ d │          │
│          │ i │          │
│   Nuc 1  │ n │   Nuc 2  │
│          │ g │          │
│          │   │          │
│          │ b │          │
│          │ o │          │
│          │ a │          │
│          │ r │          │
│          │ d │          │
└──────────┴───┴──────────┘
              Entrance
```

Above: A specially designed brood floor with two separate entrances.
Below: Frames are transferred from the brood box to two nuc boxes which are placed alongside.

Unwelcome Traits

Once you get experienced at requeening, you can use this technique with two or more hives to iron out any unwelcome traits that may have developed in your bees. It's a matter of breeding out the bad traits and breeding in the new ones. (The British Isles Bee Breeders' Association encourages beekeepers to work together on improving their colonies – address in the Appendix.)

The three most obvious bad traits that you would like to get rid of are:

1 Followers. Bees that follow you for no apparent reason are a nuisance. They will also 'follow' non-beekeepers, and when they panic and flail around

soon as you see queen cells.

Hopefully you will get a new queen in both nucs (or both sides of the dividing board), and be able to requeen both hives. If you should unfortunately only get one you should use her to take over from the one you consider to be the worst of the old queens, using the standard technique of placing the worst queen in her brood box on top with a sheet of pin-pricked newspaper between. You can then take more day old egg frames and try to make another queen to take over your second hive.

with their arms the bees will respond with a defensive sting.

2 Runners. Bees that run around the frames in a hyperactive manner will get hot and bothered when you work on the hive, and are more likely to sting than bees which continue to work the frames in a quiet, orderly manner.

However make sure you are not making them 'run' by over smoking.

3 Swarmers. Some bees just love to swarm, despite all your attempts at artificial swarming! Others colonies will carry out their own supersedure which

is a splendid trait since they effectively look after themselves year after year after year.

How to do it

Say that you have two hives and Hive A has good tendencies and Hive B has bad tendencies. If you requeen them in the normal alternate year fashion Hive A will hopefully continue to be a good hive, while Hive B will almost certainly continue to be a bad one.

The obvious answer is to requeen Hive B from Hive A. The idea is to only breed from the hive with good tendencies (Hive A), enabling you to introduce your well behaved young queen and her cohort of young workers and brood to replace the old queen and badly behaved colony in Hive B.

To do this you have to note the tendencies of the hive you like and dislike and have a firm idea which trait you want to breed out. When the queen in Hive B is in her second year, take frames with eggs from Hive A and make up a nuc in the normal requeening fashion.

At the end of season transfer the Hive A nuc of bees to a brood box and put the Hive B brood box on top with the usual sheet of newspaper in between. The young queen from the bottom box should take over. Alternatively kill the queen in Hive B and amalgamate with the Hive A nuc 24 hours later, which will ensure that the new queen takes over. Once she is established the traits of Hive B will hopefully improve.

Obviously this technique requires experience and the results are by no means guaranteed. To make proper progress you should only attempt to iron out one unwelcome trait at a time – the hive's temperament should change within 24 hours of requeening.

Making Multiple Nucs

If you have a hive with a number of queen cells you can make up a number of nucs.

For instance if you have four frames with queen cells you can make up four nucs as shown in the accompanying diagram. The queen from the hive must be removed – bump her off or use her to make a tiny nuc removed from the site – and at least one frame with queen cells, brood and stores placed in each nuc.

The nuc entrances face inwards, and the remaining hive brood frames should be distributed evenly between them. Brush off any remaining bees in even piles in front of each nuc, and remove

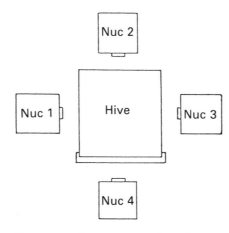

You can make up any number of nucs if you have sufficient frames with queen cells – at least one per nuc. Returning flying bees should redistribute themselves evenly.

the hive from the site.

As if by magic returning flying bees should distribute themselves between the four nucs which then should thrive and grow into four mature colonies.

Can't find the queen?

By this stage it will have become apparent how necessary it is to find the queen, and to know which one is which by marking with the appropriate year colour (page 31).

If you are one of those who can never find the queen, try this:

● Move half the frames from the brood box in question into another brood box placed alongside with its entrance overlapping. The flying bees will return to the box which contains the queen.

● Look for her again, and if you can't find her continue splitting the box until you are down to just one frame.

● Remove it from the hive area to get clear of the flying bees; if you can't find her then you need your eyes tested!

● However there is one trick left up your sleeve. Put two empty boxes together with a queen excluder between. Shake the bees onto it, and smoke them to drive them down through the rungs. Only the poor queen and drones won't be able to go.

Great care must be taken to do all this over a white sheet in case you drop her; and for goodness sake don't squash her with your clumsy feet!

Winter weakened colony

Your colony may have come through the winter in a very weak state. If the bees appear excessively apathetic when they should be bustling in the early spring they may require reviving.

On a short sleeve day when the weather is suitable, take the cover off the brood box and try sprinkling a little warm sugar syrup over the frames. This will hopefully revive them enough to be able to feed from a block of candy.

When you come to spring cleaning, if the brood does still does not seem to be growing you can try encouraging them with a weak quarter gallon feed of sugar syrup (1 pint:1lb sugar (0.56 litre: 0.45 kg)).

Don't overdo this spring feeding. If the weather turns cold the colony may retreat into a cluster leaving potential chilled brood beyond their protective area; or if the weather is good you may inadvertently over expand them into swarming.

You should of course find out why your colony is so weak, and in particular determine whether there is any disease which needs to be eradicated. If by any sad chance the colony has died out, send one of the brood frames to the National Beekeeping Unit at Luddington for analysis, and keep the hive closed until you receive their report. Remember that if it is diseased, it could be transmitted by your hive tool or other gear.

The Third Year:
The Swarming Year

As we have explained, the third year is when your bees are more likely to swarm unless you have taken the correct precautionary measures. However bees will be bees, and sometimes you just can't prevent them from swarming.

This section explains why they do it, how they do it, and what you have to do when they do it . . .

Many swarms leave the hive when the beekeeper is away, so he doesn't realise what's happened. However, if you're near, you won't miss it. Around mid-day the bees stream out of the hive, and then buzz around in a great mass well clear of the ground as they gradually orientate towards the tree or whatever they will settle on. Eyewitnesses sometimes exclaim 'the sky turned black!', which is something of an exaggeration, but you will certainly hear a loud buzzing.

Why Bees Swarm

1st year: Virgin queen.

2nd year: In her prime – you get lots of honey.

3rd year: Over the top and over-crowded – time to start swarming. They can however choose to swarm in the 1st or 2nd year.

There are very simple reasons why bees choose to swarm:
- They are overcrowded
- The queen has run out of laying space.
- As she gets older and the hive gets bigger, the queen's pheromone doesn't spread adequately through the hive. The bees it doesn't reach assume they have no queen, and start to build queen cells in preparation for swarming.

If you live in an area with a gap in the nectar flow, the bees are unlikely to swarm when nectar is on tap. However if there is no gap, as often happens with oil seed rape, the bees may still swarm.

In a June gap all the bees which the queen has produced to handle the stores suddenly find there's not enough space, nothing to do, and the pheromone is not reaching them. They start to draw out queen cells which can normally be found hanging down from the bottom bars of a brood super – usually three or four though up to 12 is possible with some in the main brood chamber.

Half the hive then decides to leave with the queen. They suck as much honey as they can to give them three days contingency rations – honey which you would have harvested – and take off into the blue yonder. Behind they leave the young queens in their cells who will regenerate the species.

What to look for

The traditional method of forestalling impending swarms is from late May to late June to look through the brood box for queen cells every 10 days – the time the workers take to form them. Finding them is not so easy. The workers cluster over them, so you will have to give a blast of your breath to clear them away. Each queen cells can then be destroyed

by pinching between finger and thumb.

This is however a bad method:
- Every time you do it, you set the hive back 24 hours.
- The bees become very bad tempered.
- They are likely to sting your neighbours.

Furthermore, as fast as you pinch them out the bees will produce queen cells and the chances are you will miss one – or in some cases end up with no queen at all, having destroyed all their attempts to make one.

The answer is to take a preventative course of action every other year from the second year onwards to guard against swarming – 2nd year, 4th year, or every year if you prefer. There are various methods of swarm control, but the two we have recommended are to requeen (page 118) or split the colony by making a nuc (page 120). Obviously you are unlikely to want to go on making up nucs indefinitely; when you have enough colonies simply requeen to induce 'artificial' rather than real swarming.

Emergency Action

What do you do if you see those queen cells in the early summer and decide your colony is about to swarm? It is a likely occurence if you have failed to use the above mentioned preventative methods.

You will have to put one or other into effect immediately.
- If you want another colony, make up a nuc using the frames with the queen cells and using the techniques described on page 120.
- If you decide it would be preferable to requeen, take out the same queen cell frames and follow the requeening technique (page 118).

It is easiest to amalgamate the new colony and old colony at the end of the season (new queen and old queen), but if the timing is right and the nuc is strong enough you can amalgamate

them in late June at the start of the main honey flow. This will give you a very strong hive under an energetic young queen which will make the most of the honey gathering. There is a possibility that the hive may reswarm, but if the honey flow is on and the hive has enough supers to give the bees room this is unlikely.

Right: This large mass of bees is a prime swarm around 25,000 strong. It initially issued from one of the author's hives at round 11.45 on a very hot day in May; unseasonably early, but at a time when there were large numbers of swarms throughout the south of England. After buzzing around for some time it went back into its hive.

Two days later it did a repeat performance, and after buzzing around for some 15 minutes gradually settled down on the tree shown which was about 20 feet from the hive, where it refused all attempts to lure it onto a frame and up into a skep. Eventually it got so fed up with this human interference that it reswarmed into a much higher tree some 50 feet away. Two ladders were needed; numerous branches had to be removed to get a clear path to it; and then it was knocked down into a skep and precariously carried down the ladder.

In the meantime the hive was moved to a new site which is a ploy to inhibit further swarming by depleting it of flying bees. The swarm was bundled into a new hive placed on the old site to catch errant flying bees, and all seemed to be well until the next day when it swarmed again!

This time the swarm settled in the willow shown hanging over the pond on page 117. It was collected by boat and hived again. About an hour later the swarm issued forth for the fifth time, and rapidly re-entered its original hive on the new site where it stayed happily for the rest of the season.

Swarms in Action

The first sign of a swarm in action is a lot of bees flying around the hive in a seemingly haphazard manner. This can happen up to 24 hours before the actual swarm.

The bees who are leaving then issue forth in a great cloud, usually around midday. The first swarm is called the 'prime swarm' – approximately 25,000 bees or half the hive following the old queen. This great black mass of buzziness is a somewhat alarming sight, although having gorged themselves on the requisite three days' supply of honey the bees are in fact in a benevolent and soporific mood. However unfortunately for the beekeeper they will have removed much of the useable honey from the hive.

Within 20 minutes the swarm should settle on a handy tree or wall, while their scouts go off to search for a new permanent home. This temporary home is usually within 50 yards of the hive, and despite being composed of different bees later swarms will usually swarm to the same spot in following years. If you know where they are going to go you can place a nuc box as a 'bait hive' there with some drawn-out frames to entice them; an easy way to catch the swarm before it goes any further. The bait hive will have to be up off the ground – in a tree, on a garage roof, etc.

The swarm hangs surrounding the queen in a great mass, and will usually stay a couple of hours before heading to its final destination. It will follow the scout bee which returns and does the most vigorous dance, and will always head for the nearest good position, though this may entail flying several miles.

The bees' final destination is likely to be much more tricky for collection – a cavity wall or a hollow tree. You should always attempt to catch a swarm at its first stopping place.

Casts

The 'casts' are the swarms which follow.

If the hive has swarmed it is ninety per cent certain that it will continue to do so at regular intervals, though this depends on the number of queen cells.

The first of the young virgin queens

A smaller cast is much more manageable to collect. Collected in May, this swarm grew into a fine colony of bees plus a honey super by late June that season.

swarms with the first cast – half of the 25,000 remaining bees; another with the second cast – half of the remaining bees again; and they may carry on and on until the final cast has no more than two or three bees, a process that happens over 10–12 days as the queens are born.

The hive is in effect abandoned, for any bees that remain will be too weak to continue with any purpose. You may not want this to happen, but it is nature's very efficient way of preventing build-up of disease by regularly changing homes.

The Role of Drones

Drones are easily recognisable. They have big bottoms with no sting; they have very large eyes meeting at the top of their foreheads; and they are slow.

When they're at home, drones can be compared to hot water bottles – with their bigger bodies they are extremely handy for keeping the brood up to temperature.

However the drones' most important role is to act as the male of the species. When they're out and about, they go round visiting hives looking for virgin queens, and are always welcome. Every now and then they fly up to join the 'drone assembly' in the sky (like a boys' day out), chase queens, mate with them, become paralysed as a result, and fall down and die!

Nor do the ones who stay behind in the hives fare any better. At the end of the season the workers have less to do and are quite capable of looking after the brood without the added complication of drones. They bite off their wings (one per drone), sting them, and throw them out of the front door.

The bees make drones to mate with queens, and if a colony is allowed to get to its third year they will make an over

Smoking can help drive the bees up onto a frame, but not too much! In the end it proved easiest to brush this swarm down into a box and unceremoniously dump them in a skep.

One danger is that a swarm might carry unwanted disease or unwanted traits – so examine them with care.

abundance of drones to aid with swarming. A well balanced colony has around 100 drones to 1,000 workers. You can tell if your colony has too many by the sight of too many drones buzzing around.

The drones mate in a ratio of about 12 to one queen. If a queen is poorly mated due to being interrupted by the weather or some other factor, she may become a drone laying queen. This means she runs out of sperm, leading to the drone laying problem which is dealt with on page 134.

Catching a Swarm

Swarm catching is a May/June occupation. The basic law of swarms is 'finders keepers', unless the original owner still has it in view – the Domesday Book stated that if you run along banging a lid and keeping a swarm in sight it's yours!

A prime swarm is a prime find since it gives you a new colony. Large casts can also be used to make a new colony, though since the new queen is unmated there is an unavoidable delay before the colony really starts thriving. Catching smaller casts is good experience, and they can sometimes be successfully amalgamated with a colony of weak bees.

The swarm you catch may be your own or may have arrived from the blue yonder – you may be summoned by a frantic phone call from a non-beekeeping neighbour. Whether catching and rehiving your own bees or someone else's the principles are the same.

Equipment & Technique

You are likely to need a skep, a white sheet, soft cloth, smoker, and two drawn out frames (taken from your hive at spring cleaning) which will entice the bees. You may also need a flat board, a pair of secateurs, a long handled pruner, step-ladder, etc.

The bees will probably swarm at midday. Once they have alighted at their temporary resting post you may have no more than two hours to catch them before they move on.

Get fully dressed and light your smoker, particularly if there is an audience present. Seeing you take such precautions they will keep well back, though in fact they are not really necessary as a swarm filled with three days' honey is very mild mannered and experienced beekeepers often catch them without any protection.

For a typical situation we will take a swarm hanging from the branch of a tree, though you can never predict

where they will choose to go:

1 Spread your sheet under the swarm, with the flat board beneath it. Hold up the skep (a cardboard box would do as

Back to our prime swarm. A fir tree like this makes swarm collection tricky. You can't brush them off or remove them hanging from a branch, but must lure them onto a frame.

well) and bang the branch to drop the bees into it; or brush them down in one quick movement; or cut the branch and lower it.

2 When most of the bees are in the skep, cover the open top with your soft cloth; then invert it, and place it down on the white sheet over where the flat board is. Push a drawn-out frame underneath to wedge it up; the drawn out wax will help to keep the bees happy in this new environment.

Leave the bees for 30 minutes. If they return to the tree you have obviously managed to leave the queen behind and will need to repeat the performance; if they stay, the queen is with them.

2a If the branch is out of reach a slightly different technique is needed. Attach a drawn-out frame to a long pole and hold it up near the bees, lodging it on a branch (or similar) which they can run along. Attracted by the wax, they will start to move onto it.

When your frame is full lower it down, and place it under the inverted skep or box with an end propping the corner up. Attach a second drawn-out frame to your pole and hold it up as before. When it is full with bees, lower it and bang the bees down onto the white sheet by the skep entrance. You may need to do this two or three times until you are confident you have got the majority of bees.

Wait 20 minutes to see if they return whence they came; if they do the queen is still up there and you must start again.

3 Knowing you have the queen, fold the sheet over the skep and pick up the whole lot on the flat board – skep, frame and bees. That way you can safely carry it some distance, or transport it in your car.

Hiving a prime swarm/large cast
A prime swarm or first cast can be recognized by its size – rough dimensions are shown in the diagram overleaf. It cannot be rehived in its old hive, but should be used to make a new colony in a new hive, or kept as a nuc until the end of the season when it can be amalgamated with another hive that needs to be requeened.

A second or third cast could also be used for these purposes, but only if you can make it strong enough.

Attempting to amalgamate small casts as soon as they are caught in order to boost weak hives normally leads to

Bees from a swarm collected on a frame and placed under a skep. You hope they rush up inside and join their queen; if not, you haven't got her.

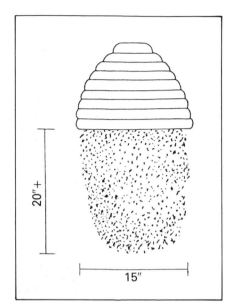

There is no mistaking a prime swarm, though the mass of bees will just about fit into a regular size skep.

their rejection.

Don't assume that you need to catch each and every cast. A large cast is a valuable free gift; a smaller cast is likely to be a waste of time.

1 The swarm or large cast will need a new hive – floor, brood box, at least two drawn-out frames plus the balance of new frames, glass quilt, spare super, spare lift if you have a WBC, and roof. Alternatively you can temporarily hive it in a nuc box.

Place your flat board with the skep on the site where the hive will be. Unwrap the white cloth so the bees can fly in and out, but construct some kind of plastic cover in case it rains. You can then leave them overnight – no longer or they will start to fill the skep with their own wax.

2 Return the next day at mid-day, fully clothed and with your smoker going. Move the skep aside and put the brood box on the hive site, ensuring that everything is done over a white sheet.

3 Leave out four frames so there is plenty of room in the middle. Gently remove the frame from under the skep which should have the bulk of bees clus-

tered on it. Place it in the brood box and then push the frames together, filling the empty space with new frames.

4 Put on the empty super. With the soft cloth over your shoulder, lower the skep down into the hive. Give it a hard bang to knock down the bees – remember to open out the flaps if you are using a cardboard box. Then quickly remove the skep and cover the bees with your soft cloth.

Put on the glass quilt and roof. The bees hate the weight of the soft cloth on top of them, so they will go down

Swarm, swarm and swarm again! A hive may continue swarming until there's nothing left. Regular requeening will hopefully ensure this doesn't happen.

into the brood box. Leave it on until the next day, when you can take off the empty super and replace the soft cloth with the glass quilt.

5 Bees will almost certainly be left in the skep. Prop it up by the entrance to the hive. If the queen is inside the hive, the rest of the bees will go and join her;

if she is in the skep, the ones inside will come out and join her.

Let them run
A more spectacular but lengthier method of hiving a swarm is to let them run in:

1 Place the frame of bees in the brood box as before. Put in the remaining frames; cover it with the glass quilt and put on the roof.

2 Use a piece of board to make a ramp up to the entrance. Cover it with the white sheet so you can clearly see the bees.

3 Invert the skep and bang the bees out at the foot of the ramp.

4 Start to smoke gently around the perimeter of the bees to drive them up. Bees always run upwards, and once the queen goes in (if she is not already in) the rest will follow. Allow 30 minutes.

Final care
The swarm should be left for three days with the entrance closed down to a few bee spaces so that it can get itself sorted out. Then you can give the bees a quarter gallon feed of sugar syrup to encourage them to start drawing out all those undrawn frames. Continue to feed them if the weather is bad.

Above: Bees from a collected swarm cluster in a thick mass both in and outside the skep. With the sheet flipped over the top and a flat board underneath, it is safe and relatively easy to carry them.

Left: A spectacular way of hiving bees! The bees are dumped en masse in front of the hive. Once the queen goes in, they follow up the ramp and into the hive behind her. Inverting the skep over the hive is a much quicker and easier method, but not so exciting! The great thing to remember is that bees that have swarmed are exceedingly mellow, and you can take a lot of liberties with them which would never be tolerated at other times of the year.

Problems: Emergency Requeening Etc

The queen may die. If the hive swarms, she may find she is unable to fly up and join them or gets gobbled up by a passing predatory bird in mid flight. Without their queen, the swarm will then return to a queenless hive, but with queen swarm cells about to hatch a new queen will hopefully take over.

However when the new queens hatch out and battle it out the wrong queen could win. She may be faulty in some way and get kicked out of the hive which is queenless again! This is extremely unlikely compared to losing a queen due to your own mishandling – dropping her and stepping on her with big boots is an unfortunate occurrence.

Whatever the cause, you must act quickly if a hive goes queenless. The symptoms are purposeless bees, no pollen being taken in, and with your curiosity aroused you find no eggs or brood when you open the hive up to have a look. If there are any queen cells you must obviously leave them alone and hope they produce their own queen; if there are none you can immediately introduce some frames with day old eggs taken from another hive so they can make queen cells.

Alternatively amalgamate them with another hive by placing one brood box on top of the other, with a sheet of pin-pricked newspaper between which gives the colonies time to adapt to their individual smells. While a weak cast is likely to be rejected, a strong colony will be incorporated. The hive can then be split once again, using the process for making a nuc (page 120) in the May-July swarming season.

Drone Laying Queen

A drone laying queen is one that is running out of semen, either due to old age or to being mated poorly. Her mating flight may have been interrupted by the weather, with the result that she mated with less than the dozen or so drones needed to fill her sac.

A queen in this state will continue to lay her eggs in worker cells, but will only fertilise them intermittently. The unfertilised eggs grow into small, squashed looking drones; the cells themselves are recognisable by the drones' dome shaped caps on worker foundation which should be on the larger cells of proper drone foundation.

Realising something is wrong – few new workers and a lot of new layabout drones – the workers may make an emergency queen cell from a fertilised egg. If so, close down the hive and leave it, checking a few days later to see if the queen has issued forth. If all goes well she should then take over the hive once she is mated.

If there are no queen cells, you must requeen. You can do this by killing the existing queen (squash her) and introducing a sheet with day old eggs from your other hive, or a frame with queen cells, or a new queen in a queen cage 24 hours after killing the old queen.

Alternatively you can amalgamate the hive with your other colony as explained above. Kill the queen first, and then put the brood box on top of the other colony's brood box, divided by a sheet of newspaper to give them time to adapt. Then wait until the early summer swarming season to split the hive once again if the timing is right.

Drone Laying Workers

In some cases when a hive becomes queenless a few of the workers (about five) will start to develop egg laying tendencies.

They are unable to be mated, so any eggs they lay will become male drones. The way they lay is noticeable, because it is haphazard with several eggs crammed into a cell – in fact the most telling sign of drone laying workers is seeing cells containing more than one day old egg.

Your hive is obviously on a hiding to nothing, but putting it right can be a problem. We recommend the following method, but the weather must be warm if your good bees are to survive:

1 Remove the hive some distance from its site, placing it on a white sheet. Replace it with a new floor and brood box, complete with two frames with queen cells or day old eggs in the centre.
2 Take off any supers. Take out the brood frames one by one, and give each a hard bang so all the bees drop onto the white sheet.
3 Put the bee-less frames in the new brood box, trying to keep them in the same order on both sides of the two frames which will hopefully produce their new queen.
4 All the flying bees should find their way back to the hive site and into the new hive.
5 If you have a brood super, follow exactly the same procedure.
6 Look to see which bees are left behind on your white sheet after 20 minutes. The drone laying workers won't be able to fly. Squash them and make sure you kill them all. They have a Rasputin-like influence on a hive, and would never allow it to return to normal. They would simply kill any new queen.

You now have a hive with no drone laying workers, and two frames with day old eggs or queen cells. Hopefully within 12 days the hive should have a new queen.

The Third Year and Beyond

Having successfully completed three years, the beekeeping cycle will continue with 'artificial swarming' being employed to keep one or two colonies at maximum productivity – the ideal colony is one where the bees never pursue you, are quiet and diligent when working on the frames, and always supersede their queens and stay happily in their hive!

In this section we come to the end of our practical beekeeping, by summarising your first four years of beekeeping with the ground rules for successfully keeping bees in ensuing years. We hope that from now you can keep a few hives with minimum fuss, maximum fun, and a lot of honey! There are plenty of more advanced techniques, but they are beyond the scope of this book and are dealt with in the wealth of specialist publications.

You may also be interested in the possibilities of commercial beekeeping, and with that in mind this section concludes by summarising the realities of commercial beekeeping and work with migratory hives.

It creeps up on you! First one hive, then two, then a few more. Once you've got a few colonies going it's easy to increase the numbers of your bees, but the cost of the equipment and the amount of time required may become prohibitive. You also need suitable sites for your colonies, and what could be better than this woodland apiary with its selection of colourful DIY brood boxes and nucs. Old railway sleepers and concrete blocks act as hive stands, and the entrances are more or less all facing in different ways to prevent drifting.

Three year's beekeeping summary

There are many variables in beekeeping as this book should have made clear, and in particular dates when various things happen will be dependent on the state of the weather as well as other matters. However if all goes well with your bees each year should follow a fairly regular pattern. There will be the inevitable ups and downs, but as you become more experienced the behaviour of your bees will become easier to predict and handle while you continue to learn about them.

The first year is for learning and making steady progress, with hopefully a few pots of honey as your reward at the end of it. The second is one of consolidation – a good crop of honey, a second hive, and the use of hive management aimed at swarm prevention. The third is more honey, more hives if you want them, and swarm prevention once again. You should have all your equipment by this stage, and as production rises the cost of each pot of honey will at last start to fall!

YEAR 1

January

Your first year of beekeeping. Evening classes on the theory of beekeeping are well worth attending if available. Join your local beekeeping association and read their regular newsletters to find out about forthcoming events.

February

If you are intending to buy a nucleus of bees from a dealer in the spring, these should be ordered in plenty of time. You should also get your beekeeping equipment well ahead of time with a basic shopping list that includes:
- Hive with brood box and super.
- Frames to fill the boxes.
- A glass quilt to go on top.
- A hive tool and smoker to work the hive.
- Protective clothing.

If you buy your hive components 'in-the-flat', you should allow time to assemble them. Alternatively if you are a committed do-it-yourself enthusiast it is relatively easy to make a single wall National hive working from British Beekeeping Association plans.

You may choose to buy your equipment secondhand. Be careful with what you buy, and allow plenty of time to sterilize and clean it up properly.

March

Make up sufficient frames for brood box and super, plus a few spare in case you need them. By now you should have decided on your hive site and have it prepared. The hive must be on level ground.

April

The hive can be installed in position ready to accept the bees. Make sure you are happy with the site as it is very difficult to move a hive once the bees are in situ. Allow yourself plenty of room to carry the hive to a secondary site for examining frames away from the flying bee.

Some beekeeping classes start to move outside at this time of year so that students can learn practical as well as theoretical beekeeping. This is extremely valuable tuition as it gives you confidence in handling bees. Ideally you should continue classes right through the season, with your teacher keeping one step ahead of your own beekeeping. If your local association has a 'Spring cleaning day', go along and watch how they do it.

May

If ordered early enough, you should take delivery of your first nucleus of bees. Hive them, feed them, and watch them grow into a full size colony.

June

Continue feeding the bees if necessary. When they have almost expanded to fill the brood box, turn the outside frames and use it as an opportunity to look through all the frames to check the queen's performance. However remember that each time you go through the brood frames you will disturb the bees and effectively set the hive back 24 hours. Do it as much as is necessary, but no more.

July

The main honey flow should be in full swing. When all the frames are covered with bees and the new wax drawn out, add a super to increase the brood area to a brood-and-a-half. This will give the queen room to increase the colony to its full size. You may be able to do this as early as June if the weather is good. On the other hand if the weather is poor foraging will be inhibited as the bees are confined to the hive, and the colony may not expand enough to require a brood super that season.

August

When the brood area is full with bees it's time to add a honey super; you may have already done so by July. This should be separated from the brood by a queen excluder, so it is for honey storage only. Remember to turn the end frames of the brood super before doing this in order to ensure that the bees draw out all the wax. The honey super will need a drawn-out frame to attract the bees up through the queen excluder.

September

Late August signals the end of the main honey flow. If you have reached the stage of putting a honey super on in your first year, you should now harvest the honey that it contains. Leave the brood honey to the bees.

There will be a lot of out-of-work workers wondering what's happened to their nectar sources – this is the time of year when most people get stung and extra care must be taken. The bees are pushing drones out of the hive entrance.

October

Decide whether to winter your hive on a brood and a half or brood box alone. The rule of thumb is that if the brood super contains brood or a substantial amount of stores, you should leave it on; if not, take it off.

Start fast feeding the bees in the last week of September. Aim to finish early in October, taking no chances on the weather turning wintery with the result that the bees can't get rid of the water content of the sugar syrup – it would then ferment and become useless to them. Be very careful to prevent robbing, by closing down the hive entrance to a few bee spaces and ensuring no drips of sugar syrup are left around the hive. Fast feeding two gallons of sugar syrup should take around three days.

November

The hive should be battened down for winter. A zinc mouse guard protects the entrance, and a wooden crown board replaces the glass quilt, supported on four match sticks to provide sufficient ventilation. Ensure the roof cannot blow off, horses or cattle cannot kick the hive, and if necessary take precautions against woodpeckers.

Clean all your equipment, and be very careful to guard against wax moth laying their eggs in your stored frames.

December

Make an occasional external check on the hive for signs of any problems, such as those caused by wind or woodpeckers. Give the bees a block of candy for Christmas! They will now be in their winter cluster, with the size of the colony diminished to some 10,000 bees. They will occasionally fly on fine days.

YEAR 2

January

The queen should start laying in January, although you probably won't be aware of it. If the bees are flying on sunny days and taking back pollen to feed the brood, then all is well in the

BEE-HOUSE TO CONTAIN TWELVE HIVES.

hive. Don't attempt to lift the crown board to have a look inside at this time of year. The propolis which holds it all together helps insulate the colony against the rigours of the winter.

If the bees make slow work of the candy, they probably don't need it. When they finish it, give them another block.

February

Time to start overhauling your equipment. Do you intend to start up another hive? If so you should get the various components and get your second hive sited in plenty of time for the new season. You will certainly need a spare floor and brood box for spring cleaning, and a supply of spare frames.

March

There should be obvious signs of pollen being taken into the hive by this stage of the season. If not, examine the brood box at the first opportunity in shirt-sleeve weather during April or May. The colony may be severely weakened and need reviving with a light sprinkling of warm sugar syrup when you can get to them.

April

If the weather is kind you should be able to spring clean, but only on a warm, still day. Check through the brood frames and note down the

amount of eggs, sealed brood, pollen and honey. If there is plenty of everything, you can look forward to lots of bees collecting lots of honey.

If you spot the queen you should take the opportunity to mark her with the appropriate colour. Before May she is likely to be the previous year's queen; any later she may be a new queen due to the hive swarming or superceding unbeknown to you.

Put in two new brood frames marked with the colour for the year, starting a rotational system of regular replacement. The two frames taken out will come in useful for catching or hiving swarms.

May

The brood area should be a brood and a half by now, and the first honey super in place if you live in an area of oil seed rape. The rape honey flow can extend from mid-April to mid-June depending on the weather and when it was planted.

Rape honey crystalises very rapidly, and must be spun-off promptly. The bees can fill a super with rape honey in four to 10 days. It is best removed once it has thickened enough not to drip from the frames, but before it is properly capped off. Any later and you can only cut it out.

The quantity of rape honey you are likely to get is variable; in his second year the author bottled 42lbs taken from two supers.

May is also the time to prepare for swarms. The old queen and half the bees which are predominantly your flying bees may leave the hive, putting a severe dent in your honey production. Apart from making sure the colony is not overcrowded, the traditional way to combat swarming is to look through brood frames every 10 days, and pinch out the queen cells. However regular requeening is better practice.

June

At the end of the rape flow the bees may get short tempered for a short while; watch out for stings! Some areas will have a 'June gap' during which the bees cannot subsist on the amount of available forage. If this appears to be the case, feed them a weak mixture of sugar syrup until the main honey flow commences. They won't touch it if they don't need it.

This is the time when you should start the requeening process, by taking out brood frames with day old eggs to produce a new queen with which to requeen the hive later in the season. Alternatively you can create a new colony for a second hive by splitting the existing colony to make a nucleus.

Both methods will hopefully prevent swarming, but this is by no means guaranteed. Running two hives is no more difficult or time-consuming than running one and makes overall management of your bees much more easy; it is also sensible 'insurance' in case one colony dies or swarms out.

July

The main honey flow usually runs from July through August. Keep an eye on the bees' progress, giving them enough honey supers for their requirements. If the weather is bad, be prepared to emergency feed them with sugar syrup.

August

If you have produced a new queen for requeening, introduce her by amalgamating the two brood boxes – old on top of new. She should kill the old

queen, giving your colony a new lease of life.

Take the honey from all honey supers by the end of August, taking great care as the bees will be very bad tempered. Put the wet supers back on the hive so the bees can clean them up. From now on any honey they collect is their own.

September

Start fast feeding in preparation for winter. Remove the brood super if it does not contain brood or sufficient stores.

October

Prepare to winter down as in your first year. Clear away any vegetation which might lead to dampness in the hive.

November

The hive or hives are wintered down, with the bees only flying on fine days.

December

Give your friends and neighbours a jar of your honey for Christmas. Give the bees a block of candy.

YEAR 3 AND BEYOND

January to March

Feed candy, monitor your hives, and prepare for the season as in previous years. Ensure you are not caught short with too few frames or supers in the busy months to come.

If there appear to be problems with your hive it may be due to disease. Send suitable samples of bees, frames or floor debris for analysis to eradicate the problem and ensure it is not serious.

April to June

Spring clean your hives, marking queens and checking the contents and state of the brood frames. Introduce two new brood frames on a rotational basis each year, with colour coded ends so you can keep track of them.

Build up the colony to cope with the rape flow, removing rape honey in plenty of time before it solidifies.

Hives with second year queens should

be requeened as a matter of regular anti-swarming management; or if you want more hives you can split the colonies to make sufficient nuclei. If you spot queen 'swarm' cells this should be done in a hurry, but however good your management you will inevitably lose some swarms.

A skep will be a useful accessory for catching any swarms, whether your own or from other sources, and a nuc box is useful for hiving them.

July to September

The bees will collect the main honey flow, and you will steal it from them by late August. Then fast feed the hive preparatory to wintering down.

If requeening, the new queens should be introduced in August.

October to December

Winter down, store away your equipment, and think ahead to the following year. By now you are hopefully getting large honey harvests, and have enough excess wax to try your hand at making candles, wax and polish. For these you will need some kind of wax extractor.

You may care to enter your honey for competitions culminating in a prize at the National Honey Show in October, or use it to brew your own mead to help that wintertime reflection.

The continuation of the species — queen cells allow the colony to continue with a new queen, or divide and swarm

MORE TROUBLE-SHOOTING FOR BEEKEEPERS

Q. I have looked in the brood box and found queen cells in my first year of beekeeping. What does this mean?
A. Problems like this in your first year with a new nuc of bees are frequently caused by interfering with the bees too much – opening them up, pulling the frames around, fussing over them, etc. Leave them in peace, or they may decide to head off for a new home.

If there are a lot of queen cells the bees are likely to swarm. This is an ideal opportunity to commit yourself to a second hive by splitting the colony to make a nuc. If there are a few, they may be superseding the queen which implies that the present one is old or has something wrong. To be on the safe side it is probably best to make an 'artificial swarm'; split the colony to make a new

queen, and then reamalgamate the bees. This can be done at the end of June to make the most of the main honey flow; or later in the season, right up until the time you winter down.

Q. My bees are dying in front of the hive. What's wrong?
A. Drones will be thrown out before the autumn, and you may occasionally see old workers dead through natural causes outside the hive.

However if there are a lot of dead bees, the reason is most likely to be the effects of toxic spray. If you live in a farming area this may be due to farm spraying: more likely it is your neighbour spraying a weedkiller or insecticide in his garden. Try and find the source of the problem.

Q. My frames disintegrate when I attempt to spin the honey off. Why?
A. Start spinning slowly, and

gradually build up speed. Old frames tend to be stronger than new ones, and obviously you must never attempt to spin-off unwired frames. If the weather is very hot, the wax will soften and make disintegration more of a possibility.

Q. My hive is showing no signs of activity in December. Is anything wrong?
A. Probably not, though you should worry if there is no sign of the bees flying and collecting pollen in February/March. At this stage of the winter they should still have plenty of stores, but give them a block of candy for Christmas in case they require it. If you are certain that the bees are starving, sprinkle some blood heat sugar syrup through the feed hole.

Whatever you do, don't open up the hive to take a look inside until you spring clean in suitable weather.

Commercial beekeeping

Most of the honey on sale in our shops is foreign honey. It's an unfortunate fact that the UK spring and summer is too short lived to compare with the beekeeping climates of major commercial producers such as New Zealand, Mexico, the USA and South Africa.

Nevertheless there are commercial beekeepers in the UK, mostly working on a fairly small scale. Few, except those who have access to prized heather honey on high ground, will reckon on making money out of honey sales because the figures just don't add up to healthy profits.

For instance if you have 100 hives producing 30lb (13.60kg) each of honey, you have a total of 3,000lb (1360.8kg) for sale. Sell it at £1.50 per lb and you have £4,500 for a year's work minus the retailer's mark-up; minus all your sugar for feeding; minus the cost of replacement gear. That leaves you with a miserly wage to pay yourself and for all the other overheads, and if there's a year when you get virtually no honey as does happen in the UK, the net result is that you get no money!

The only way to make money out of honey is to have a very large number of hives which is the capital intensive formula used by the top Scots' heather honey producers who are farming a unique and highly prized honey. For other beekeepers there is no way they can compete with beekeeping countries that are blessed with better bee climates, and even with their better weather the foreign bee workers have to work exceptionally hard to make modest incomes.

Those who do make a living out of beekeeping in the UK combine income from honey sales with income from pollination fees, and also from queen rearing which is a thriving branch of commercial beekeeping due to the ban on imported queens from so many countries. Some sell beekeeping equipment, while other income can come from selling wax, teaching, etc.

The Commercial Beekeeper's Year

This book is not a guide to commercial beekeeping which demands a great deal of experience as well as commercial acumen. However basic details of the life of the commercial beekeeper are interesting.

The commercial beekeeper's year runs along much the same principles as the hobbyist beekeeper's year. Only everything works on a much larger scale and is consequently a lot more complicated, involving much more time and physical work – unlike farmers who use tractors and other sophisticated machinery to replace manpower, there are few short cuts in beekeeping.

Seven nucs of bees being raised for the season ahead. On this scale the amateur becomes a serious amateur, but it's still a long step to the full professionalism of commercial beekeeping and its attendant problems.

At the start of the year the commercial beekeeper checks that all his equipment is in order and he has enough of what will be needed – frames, supers, brood boxes, nuc boxes, etc by the dozen! Most favour larger Langstroth or commercial size National hives since they have greater capacity than domestic size hives.

The commercial beekeeper has the

same spring cleaning requirements, but with so many hives this is usually impossible to achieve on a yearly basis. He then adds supers as his colonies build up towards the main honey flow, and finally harvests the honey. Putting supers on and talking them off demands physical fitness when you're dealing with dozens of hives.

Traditionalists use clearing boards to drive the bees down when they want to take the honey, though chemicals which are available such as Benzeldahyde make the process much faster. However great care must be taken not to overdo its use which can drive the bees straight out of the hive, and will inevitably anger them.

The full supers are transfered to a closed purpose-built trailer or lorry, on which they slot into trays to ensure there is no dripping. They can then be transported back to base for extraction, settling and bottling.

Obviously as with the amateur, oil seed rape makes the commercial beekeeper's job harder, though the quantity of honey produced is greater. With the threat of rapid granulation he has to keep checking his dozens of hives at weekly intervals, removing the supers

and immediately replacing them with new supers full of drawn-out frames – this demands a large stock of extras. The main honey flow of a non rape season is easier for the professional as it is for the amateur – each hive can have three supers piled on top, and then the whole lot are cleared at once.

When it comes to wintering down, all the hives have to be fast fed in the usual fashion, though in big open areas there are less robbing problems than the hobbyist has in a small suburban garden. Surplus supers are removed, and all hives are left in situ except for the migratory hives which are moved to a winter apiary.

The main winter dangers are likely to come from mice, woodpeckers and cattle. Mouse guards deal with the first problem; hives are covered with chicken wire or black plastic to forestall the second; and the third can only be checked by agreement with neighbouring landowners.

3 Year Cycle

Like hobbyist beekeepers, the commercial beekeeper also relies on requeening to ensure maximum productivity for his hives, and help guard against swarming.

Ideally every colony is requeened or split down annually, though with so many hives it may be impossible to manage this.

More than anyone the professional beekeeper relies on having very strong colonies, and keeping them strong. The amateur may lose a couple of colonies which is sad; the professional may lose a couple of dozen which is a disaster.

Stony Meadow has space for some 60 hives at the height of the season, plus many more hives at out-apiaries.

143

Migratory beekeeping

Real enthusiasts with over 40 hives can go semi-professional by joining the Bee Farmers Association (address in Appendices) This is a UK scheme whereby farmers 'book' hives to pollinate their crops. It is an important natural aid to the farmers – it is estimated that bee pollination can almost double top fruit crops such as pears and plums.

The Bee Farmers Association have a pollination list and tell their members when and where to take their hives, to fit in with the main pollinating times. The association collects a fee from the farmer, and then passes on payment to the beekeeper. This payment should take into account the fact that colonies used for pollination can seldom also be used for honey production, and they and their hives may well have a hard life.

You may wonder why farmers don't save money and keep their own bees for pollination. The problem is that they have no way of telling their bees what crop they should be working on, and they will invariably go out and work the wrong one.

For instance a farmer may buy some hives and site them in his orchard, but the bees may find an alternative source of nectar and pollen (in particular oil seed rape) which comes on stream before his fruit blossom is ready. Once they go to work on a crop the bees will work it until the end, and in the meantime his fruit bloom may flower and fall while the bees ignore it.

The farmer can only solve this by moving in migratory hives at the moment his fruit blossom is ready. Despite its comparatively undersirable low sugar content, the newly arrived bees will go to a crop which is immediately available for foraging on their doorstep, and won't bother to discover a field of rape which is further away.

The same principles must be used with other crops. If the bees are needed to pollinate beans and there's a field of

rape nearby, the hives must be taken right into the middle of the flowering beans so the bees have only to go a few inches for their nectar and pollen supplies. Bees like a clear cut decision, not a choice!

Moving Hives

Managing migratory hives is hard work.

Rather than loading and unloading them at each new site, most professionals use specially built trailers which can be towed to the required spot late in the evening with entrances closed; and then left there with entrances open for the bees to forage until the flowering of that particular crop is finished.

The beekeeper then closes up the entrances (late in the evening again), hitches up, and drives his trailer of hives to its next booking, with a possible stop to change supers, spin-off honey, etc.

One problem is the threat of theft or vandalism. Despite its tranquil image, beekeeping does attract a small amount of crime. 'Bee rustling' is a somewhat

Above: Broad beans are a favourite new crop for migratory hives which must be timed to take full advantage of their flowering and pollination.

specialised form of theft which nevertheless takes places, but on a more mundane level trailers are prized possessions.

In one notable case in Sussex a open trailer of hives was left in the middle of a bean field to pollinate the crop. Someone decided to steal the trailer, but did not want the bees. He therefore put a rope around the hives, attached it to a stake driven into the ground behind the trailer, and then hitched up the trailer to a car and drove off. The obvious result was that the appalled beekeeper returned to find no trailer, and a pile of broken hives and destroyed colonies littered over the ground. Where possible, lock your migratory trailer!

As an alternative the hives are likely

to be more effective scattered throughout the crop, though this of course involves heavy unloading and loading. The bees must also be encouraged to collect pollen, which is done by ensuring that they have large brood areas which need the pollen for food; by feeding with sugar syrup which turns nectar gatherers to pollen gatherers; and by providing water nearby so they don't have to waste valuable time collecting it.

Obviously the beekeeper must ensure that no insecticides are applied when the honeybees are working the crop

Understanding pollination

Pollination is the transfer of male pollen from the anthers to the female stigma of the same or another flower of the same species.

Some plants will self-pollinate; for many pollination by wind, birds or insects is desirable, or in some cases a necessity. Bees of all kinds are the most effective external agents. The branched hairs of their legs are designed to carry pollen to feed their brood, and their pollinating prowess as they brush from flower to flower can be used to improve the quantity and/or quality of a crop whether they are collecting pollen or nectar.

The situation today is that intensive farming and the use of insecticides and herbicides has so reduced wild bees that domestic honeybees worked on a migratory system are becoming more and more important, since they are the only insect that can be 'worked' by man.

Often a vast crop will only sustain visiting bees for a short period of time; then the bees must be moved to find sufficient forage for their colonies.

The number of colonies required to pollinate a crop is a variable which depends on the pollination needs of the crop, its size, its flower density, the amount of forage for the bees, the number of competing pollinators, and the effectiveness of honeybees on that particular crop. Two or three colonies per hectare is a rough and ready average for most crops, with local variations.

Beware the modern honey rustler! Six hives packed with rape honey and an expensive trailer may tempt the criminal minded — so lock it!

The Honeybee's Products

Keeping bees is a pleasure – it's stress reducing; it's an excellent and absorbing hobby; and you will keep on learning about the bees year after year. However when it comes down to it, most of us keep bees because of what they produce for us. Number one is honey which comes in as many tastes and textures as the world of wine. You can harvest it and enjoy it for your family and friends; you can show it and win prizes; if you have enough hives you can sell it; you can use it in all kinds of recipes; and you can also use it to make your own mead.

Bees also produce wax, and you can use that to make the best and most heavenly scented furniture polish and candles that money can buy! The propolis which bees produce is also believed to have some very special health giving properties – these and all the honeybee's products are summarised on the following pages.

All the good things that the bees bring us! Honey is of course number one, and as an alternative what about a fruity mixture such as honey & apricot, honey & kiwifruit, or honey & boysenberry? There are also wax products such as magnificent beeswax candles (rolled foundation candles are shown) and perfect beeswax polish. And then of course there are all kinds of pollen, propolis and royal jelly products to appeal to the healthy minded. . . .

What is honey?

Honey (huni) is a sweet sticky fluid made by bees from nectar; (fig) anything soothing; (coll) darling, anything delightful.

The above is a typical dictionary definition of honey, and it will probably go on to be equally flattering about the various honey word derivatives: *honeyed* (fig) flattering, coaxing; *honeymoon* (fig) period of untroubled happiness! The word itself is derived from the Arabic 'han' (that's what they call it), which the Germans changed to 'honig' and our English ancestors to 'hunig'.

What it doen't mention is that honey has a delighful variety of tastes which are dependent on the nectar sources and the soils of the growing area. When properly stored it is also immortal – the contents of a sealed honeypot in an ancient Egytian tomb proved to be completely edible!

Constituents

On a more pedantic, unpoetic level, an average sample of honey would contain 38 per cent frustose (also known as levulose), 17 per cent water, 31 per cent glucose (also knows as dectrose), and 2 per cent sucrose plus minute amounts of other sugars, oils and enzymes.

It is a predominantly energy giving carbohydrate food. The amount of protein is very small, and if the honey has been filtered to give a clear appearance the protein may be reduced to nil. Natural honey also contains small amounts of vitamins – thiamin, ascorbic acid, riboflavin, pantygiothenic acid, rydoxine and niacin. Filtering may remove them also, and overheating the honey for any reason will destroy them.

Nectar becomes honey when the bees add their enzymes causing the higher sugars to invert, thereby creating the product we love in an entirely natural way. Glucose sometimes predominates over fructose in which case the honey will granulate more quickly as it turns into sugary crystals – as we have seen,

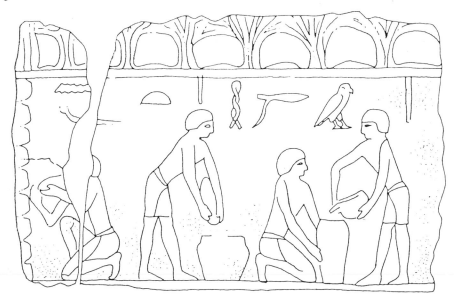

The ancient Egyptians liked beekeeping too! These chaps can be seen on the wall of a tomb taking out honey combs and packing them in jars c. 2400 BC!

this is particularly true of oil seed rape honey.

The good news is that man so far has been unable to make artificial honey. The sticky substances which the scientists have produced may look the same and taste vaguely similar, but they have not managed to find the secrets of honey's health giving properties.

Experiments have been conducted on these health giving aspects, and it is generally agreed that honey is a fine food for promoting energy, good blood condition and a healthy appearance.

Honey Fermentation

Foragers only visit plants with nectar sugars in excess of 20 per cent in relation to water content. More water and it's just not worth the work involved in getting rid of it. Bees visit crops with a high water content such as pears and plums mainly for their pollen.

When the honey is no more than 17 per cent water, the bees will seal it over with wax for storage. More water would allow honey to ferment with natural yeasts which are always present in the air, creating an unpleasant alcoholic solution which is equally unpalatable

for man and bees, resembling and smelling like old treacle. In a very wet summer wild yeast can be transported by the bodies of the bees, making their honey very prone to fermentation.

Uncapped honey should not be spun-off because sooner or later it will ferment – rape honey is the exception. If for any reason it is uncapped, you should give it back to the bees or consume it within a short time.

Fermentation can also occur if the beekeeper is careless. Apart from ensuring all honey is ripe, take care that no water gets into it (wet jars etc) and try to regularly sterilise your equipment with boiling water. So long as you keep it reasonably clean you should keep clear of fermentation. Bottle your honey in stone or earthenware jars and it should keep as long as that Egyptian; tin or metal containers are fine in the short run, but acids from the nectar-providing plants will penetrate metals over a long period of time.

Honey for sale

Honey can be produced for home consumption or sale in a variety of forms. The beekeeper can primarily choose liquid honey, cut comb, or honey sections.

Liquid Honey

This is the natural product spun straight off the frames, which though it starts 'thin, will inevitably become 'thick, in time.

All good honey crystalises and becomes thick due to its sugar content. Oil seed rape honey cystalises in a matter of days; other honeys may not crystalise for several months; top fruits such as pears which have a low sugar content will remain liquid a long time. You can temporarily liquify thick honey by heating it, but if you overdo it there is a danger of killing its health giving properties and relegating it to being yet another sweet, sticky substance.

'Creamed honey' is a specialist and currently popular application made by beating partially liquified honey to produce a soft condition. To do this the honey is carefully warmed and then beaten with a plunger. When it is bottled it settles and recrystalises, maintaining a cream like consistency for easy spreading. Great care must be taken not to overheat the honey in the first place; and not to allow the plunger to break the surface, letting in air and leading to possible fermentation.

Cut Comb

Cut comb honey is produced with special unwired thin foundation. When the wax is drawn out, filled with honey and capped, a special cutter cuts out pieces which can be stored in jars or

You can get 30 lbs of honey from the frames in one honey super. If you want to sell any excess, it must be labelled to suit the regulations.

149

plastic containers. A refinement which is commercially popular is to pour liquid honey around the comb; this makes up the advertised weight on the jar.

With cut comb you get the whole lot – honey, wax, pollen and all the bits the bees have left behind. The thin foundation ensures the wax is not too predominant, and the overall taste is extra healthy.

The obvious drawback is that once the comb is cut the foundation is destroyed, and you have no more drawn out wax. Your bees the following season will have to build wax in addition to making honey, making them less productive than their liquid honey rivals.

Sections

Section honey is without a doubt the most attractive way to present it, and attracts a high price commercially! In essence it is the same as cut comb, but each comb is contained within its own box frame, three of which are equivalent to a conventional super frame.

Most beekeepers who produce honey sections use complete 'section racks' which are placed over the brood box instead of supers. The bees don't like them, and will swarm unless a specific technique is used to persuade them to go up on the sections.

Section technique

The beginner can only work sections successfully with a swarm; otherwise the result will be that your bees will swarm!

Take a prime swarm and hive it in a brood box filled with drawn-out frames so the queen can start laying. Feed it with sugar syrup (1lb:1pt [0.45 kg: 0.56 litre]) for 10 days, and then place your section rack over the brood box with a queen excluder in between. Don't feed any longer, or the bees will start to store syrup instead of honey.

The workers won't like it, but the queen will not let them re-swarm as she is happy laying in the drawn-out frames. Filled with sugar syrup the workers have their wax glands stimulated into working, and have no choice but to go up and draw out those nasty sections.

Productivity of cut comb is a chancy affair; also like cut comb all the drawn-out wax is sold with the honey, plus the frame as well! You are therefore back to square one for honey production the next season.

The more hives you have, the larger the extractor you need for this is a time consuming and messy buisness. A large modern stainless steel or plastic radial extractor is expensive, but the old galvanised style shown would not pass EEC regulations.

Commercial Requirements

For the non-beekeeper, there is nothing like buying home produced honey. He can choose from bottles and bottles on supermarket shelves, but few apart from the very expensive brands will compete. Most will be labelled 'Produce of more than one country', and that is why they taste so poor. They contain a mish-mash of honeys from different countries 'blended' according to economic dictates, and the subtle taste that characterizes the best honey is lost amid an overbearingly sweet but flavourless result.

You are unlikely to take on the supermarkets, but you may start to sell a little or even a lot. Two hives may give you enough excess honey beyond your own needs; 10 hives certainly should. However if you take money for your honey, you must by law conform to EEC regulations regarding extraction and labelling. These are likely to change along with the vagaries of EEC bureaucracy, and you should contact your local Trading Standards Officer for information on the current regulations.

However the basic principles are likely to remain the same:

Extracting. Extractors must be plastic or stainless steel. The powers-that-be claim that galvanized steel can effect the honey. You wouldn't know it, and the old style tin plate extractors are perfectly acceptable for home consumption.

Bottling. Honey should be filtered through damp muslin or similar material to extract all the bits and pieces. Unless two-thirds of each frame is sealed extracted honey must be put in a ripener for 24 hours so that it can settle. This allows any froth and air bubbles to rise to the surface and be discarded. You may then have to warm it slightly to transfer it to bottles.

Labelling. Labels must state where the honey is produced (area and country); give the name and address or phone number of the producer; and state the net weight. These requirements may change

Pricing. Thankfully there are not any regulations on pricing! So long as the customer agrees to pay, you can charge what you want!

Never forget where it all comes from, whether you're producing a few pots of honey for your own consumption or thousands for sale throughout the land.

Honey for show

Honey Shows tend to take place at the end of the season around October time when the year's honey has been harvested. They are primarily organised by local beekeeping organisations, plus the BBKA National Honey Show which takes place in London each year (see page 182 of Appendices). If a local show gets over 100 entries, it receives a BBKA blue riband which gives the organisers an incentive to promote it well.

Other honey shows are organised to coincide with important events, such as the South of England Show held at Ardingly (Sussex) each spring, which apart from farming, horses, and all manner of stalls has a large marquee devoted to bees with a central table groaning with prize winning honey products.

The Classes

There are numerous competition classes at a honey show. For instance:
Clear medium honey.
Clear dark honey.
Crystalised light honey.
Crystalised dark honey.
Honey sections.
Honey frame.
Beeswax.
Mead.
Honey Cake.
Beeswax Candles.

The glory comes in taking the winners' cup, knowing you have produced the almost perfect honey, section, frame, etc for that year. The work that goes into producing these masterpieces can be intense, and at first you may be disappointed. The only way you can succeed is by experience; anyone can enter – ask your local association for details – and the sooner you do so the sooner you will learn.

The Judging

Honey show judges are few and far between. It is a highly skilled job, requiring very advanced BBKA exams to qualify. At present there are 20 or so judges in the UK.

There is one judge for the show, ideally brought in from outside the area. He is paid a fee plus expenses, and is put up and suitably looked after. When confronted by row after row of gleaming pots of honey, his first priority will be to get the numbers down to a workable minority – say six. Thus every entry with a blemish will rapidly be rejected as he scans them for dirt on the jars, specs in the honey, wonky labels, etc.

Having arrived at the final half dozen, he will taste, smell (quickly, for the honey soon loses its volatile aroma when the lid is removed), and also ensure that there is a precise 16oz of honey in a 1lb jar. This may take half an hour of intense concentration, and will obviously rely on his extensive experience.

Finally he will arrive at 1st, 2nd and 3rd places. Are there any disagreements? There may be disappointments, but generally beekeepers are extremely friendly and will bear no grudge if they don't win!

Preparation

If you want to enter a jar or two of honey in a show, put by the ones that look particularly good at the time when you're bottling so that you can make a final selection nearer the time.

Ensure that the jars are absolutely clean, and that they are filled as close to an exact 16oz as is possible. The honey must be sieved through a fine muslin or nylon filter to remove the pollen and any detritus which may be a healthfood bonus but is not a winner in the eyes of judges! They want clear honey to be absolutely clear, and will hold shortlisted jars against a torch light to check on this. If it won't filter it will need gentle heating – standing a honey tin in warm water will usually suffice. Apart from the main honey label the jar must be anonymous; in the show

each entry will be given an identification number by the organisers.

All good honey will crystalise in time after it is bottled, so if you want to enter one of the clear honey classes you will have to liquify it again, gradually bringing it up to heat. A simple method is to stand the honey on a storage heater; alternatively you can make your own 'hay box' with polystyrene and plywood heated by two 25 watt light bulbs which should liquify crystalised honey in around three days.

A more basic method is to warm it up in the oven. It may not be widely remembered that as well as founding the Boy Scouts Baden-Powell was an enthusiastic beekeeper and honey show cup hunter between 1925 and 1940. West Sussex veteran beekeeper Arthur Chitty tells the anecdote of how BP left his honey in the oven; his wife turned the heat up not realising there was any honey there; and the ensuing extra dark honey won prizes for many years after!

BP also won prizes for many years with his beeswax, for the same entry can be reused year after year. The obvious exceptions are candles; mead which needs to be tasted and should therefore be entered with a flanged stopper; and the honey cake which is baked to a recipe set by the association

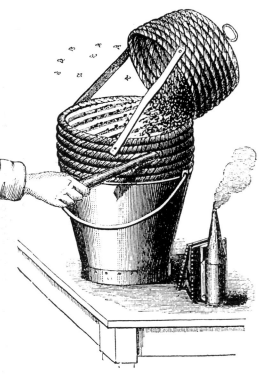

Exhibits at the National Honey Show in the Schools Class.

and must also be tasted. Colonel Bartlett – a great honey cake prize winner – recommended 'shut the old woman out of the kitchen'.

The matter of honey taste is more complicated. It will of course depend on where your bees go to forage, and as we all know you can't give them precise instructions in this matter. If you have

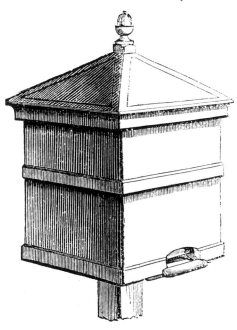

the means, just take them to the right place – red clover, white clover, heather, acacia and broad bean honey all have particularly good flavours – and hope.

Preparing crystalised honey is obviously a simpler matter, and with oil seed rape honey there are only very small differences in taste. Take care to scrape any pollen specs off the surface. If there seem to be unsightly streaks running down the sides of the honey, well and good – these are natural air spaces which the judge will look for in good crystalised honey.

Producing a perfect drawn and capped frame full of honey is more a matter of luck. The judge is looking for an unblemished appearance which can be uncapped with one stroke of a sharp knife. Arthur Chitty, who last remembers 'producing a perfect frame in the 1920's', recommends a very strong honey flow, a very strong colony of bees, and being able to take the super on and off within a week as the initial requirements.

Preparing Beeswax
Preparing a perfect 8oz block of beeswax is probably the greatest chal-

lenge for the honey show enthusiast. The experts agree that you can spend hours and hours, and then everything goes wrong.

The wax should be strained through the finest stocking into a glass dish (wash with a little Fairy Liquid first) sitting in a bain marie with water of the same temperature surrounding it. It must then be left to set perfectly with no ripples, but any movement such as a heavy footfall or passing train will disturb it. Leave it overnight with the top flooded with a thin layer of water, and in the morning you should find the perfect set block floating in its dish – and then it cracks!

The candle classes made from beeswax are dipped, rolled and moulded. Shortlisted candles will be lit, so they can't be shown again. This is necessary to show that the wax is properly adhered to the wick rather than being just a hollow tube, and that the candle burns as pure beeswax.

Overall 'cup hunting' at honey shows is hard work – it is the quest for perfection!

Peel the oranges, chop the flesh roughly, and put into a food processor. Add the honey and yoghurt and process until all ingredients are blended.

Pour into individual serving glasses and place in the refrigerator for at least two hours before serving. You can sprinkle the top with chopped nuts, or add any fruits.

Honey & Apricot Yoghurt
½pt yoghurt
2 tablespoons clear honey
Juice of half an orange
4oz dried, chopped apricots

Place all the ingredients in a bowl, cover, and leave overnight in the refrigerator. To serve, spoon into individual glasses.

Honey Baked Apples
1 tablespoon honey
4 cooking apples
4oz ground almonds
2 tablespoons orange juice
Grated rind of 1 lemon

Core the apples. Mix together the almonds, lemon rind, orange juice and honey, using the mixture to stuff the apple centres.

Place the apples in a dish with a little cold water. Bake for around 45 minutes at 180°C/350°F until soft.

Serve with cream or yoghurt.

Honey & Cream Cheese Pie
4 oz flour
2oz butter
3 tablespoons of cold water
1 tablespoon of honey
2 eggs
6oz cream cheese
4fl oz milk
4oz white sugar
Grated peel from 1 lemon
1 teaspoon cinnamon

Combine the first three ingredients to make a light dough, and spread onto a removable base flan tin.

Beat the cream cheese and milk together, and then stir in the remaining ingredients. Pour into a pastry case and bake in a preheated oven for 45 minutes at 180°C/350°F. Check to see if the top is getting too brown; if it is, cover with a sheet of buttered greaseproof paper.

Pears in Red Wine
6 ripe pears
1 tablespoon lemon juice
4 fl oz dry red wine
8 fl oz clear honey
1 cinnamon stick
Grated rind of half a lemon.

Peel the pears and cut in half lengthways. Scoop out the cores with a teaspoon. Place the pears cut-side down in a large shallow baking dish in a single layer, and sprinkle with lemon juice.

Combine the remaining ingredients in a pan and bring to the boil stirring occasionally. Pour the syrup over the pears, cover the dish with foil, and bake at 180°/375°F, basting occasionally for approx 20 minutes or until tender. Remove cinnamon stick before serving. Serve chilled.

CONFECTIONERY
Caramels
1lb honey
1lb brown sugar
9fl oz evaporated milk
4oz butter

Put honey, sugar and a pinch of salt in a thick enamel pan, and bring rapidly to the boil. Stir occasionally as you gradually add the butter and milk, bringing the mixture back to 120°C.

Keep stirring to prevent the mixture sticking, and then pour the mixture into a flat, well buttered pan. Allow it to cool, and then cut into squares.

The caramels are best stored in the fridge and are good served as a treat with after dinner coffee.

Chocolate Honey Fudge
1lb Demerara sugar
2oz honey
½pt milk
5oz butter
3oz good quality plain cooking
  chocolate

Put all the ingredients together in a large, heavy base saucepan. Place over a low heat, stirring all the time until the sugar has dissolved. Then bring to the boil, and boil until the mixture has reached 115°C which will be marked as 'soft boil' on a sugar thermometer.

Remove the pan from the heat immediately, and leave for 5 minutes. Then beat it with a wooden spoon until it becomes thick and crystals begin to form. Pour into a greased 6 or 7in square tin or similar container, and cut out when set.

Honey, Raisin & Walnut Fudge
1lb Demerara sugar
2oz honey
½pt milk
5oz butter
2oz chopped walnuts
2oz seedless raisins

Follow the recipe above. For variation you can substitute brazil nuts and sultanas.

Panaforte
5oz clear honey
4oz Demerara sugar
3oz chopped pine kernels
3oz chopped walnuts
3oz blanched chopped almonds
3oz chopped Angelica
3oz chopped pressed dates
2oz cocoa powder
½ teaspoon ground mace
½ teaspoon ground coriander
¾ teaspoon mixed spice

An Italian speciality. Put the honey and sugar into a medium size pan, and place on a gentle heat until the sugar dissolves. Bring to the boil until the temperature reaches 115°C using a sugar thermometer.

Remove from the heat and stir in the rest of the ingredients. Press the mixture into a loose bottomed 9in cake tin or flan ring, and bake at 150°C/300°F for 30 minutes.

Turn out and allow to cool on a wire tray. Serve with coffee, cut into slices.

CAKES & BISCUITS
Honey Fruit Loaf
1 tablespoon honey
3oz brown sugar
2 eggs
3oz butter
4oz wholemeal flour
4oz self-raising flour
1 teaspoon baking powder
1 teaspoon mixed spice
8oz mixed fruit, nuts, chopped peel
Pinch of salt

Cream together the honey, sugar, eggs, and butter, and combine with the rest of the ingredients.

Pour into a greased loaf tin and bake at 180°C/375°F for 50 minutes. Turn out, but keep at least a week before cutting. Slice thinly spread with butter.

Honey Tea Brack
2 tablespoons thick honey
½lb mixed dried fruit
¼pt strained cold tea
1 beaten egg
1oz butter
3 tablespoons milk
8oz wholemeal self-raising flour

Put the dried fruit, honey and tea into a bowl, stir, cover, and leave overnight to soak.

Beat in the remaining ingredients and put into a 1lb loaf tin. Bake in the oven at 170°C/325°F for 1 hour or until a skewer comes out clean when inserted into the middle.

Honey Biscuits
1 tablespoon honey
10oz plain flour
4oz sugar
2oz butter
2 eggs

Rub the butter into the flour and sugar. Add the eggs and knead into dough. Roll out very thinly, and cut into biscuit shape rounds. Brush the rounds with liquid honey and bake on a greased sheet in a preheated oven at 180°C/375°F for about 10 minutes. Let the biscuits cool on a wire rack.

Yoghurt and Honey Wholemeal Scones
8 oz wholemeal self-raising flour
2 oz butter or margarine
3 tablespoons clear honey
6 fl oz yoghurt
Milk to glaze
Demerara sugar

Place the flour in a bowl. Add the fat cut into small pieces and rub into the flour with your fingertips until the mixture resembles fine breadcrumbs. Add the honey and yoghurt and mix lightly with a fork to form a soft dough. Turn onto a lightly floured board and knead lightly. Roll out the dough to an 8 inch round, and cut into eight wedges. Place the scones onto a baking sheet, brush with milk, and sprinkle with a little demerara sugar. Bake for 12–15 minutes at 220C/425°F

Honey and Yogurt Cake
4 eggs
8 oz caster sugar
4 oz melted butter
¹/₃ pt plain unsweetened yoghurt
¼ teaspoon bicarbonate of soda
10 oz wholemeal plain flour
3 teaspoons baking powder
4 oz clear honey
4 tablespoons water
Strip of lemon rind
Cinamon stick
2 oz toasted flaked almonds

Grease a 13 inch x 9 inch Swiss Roll tin. Separate the eggs. Place the egg yolks and the sugar in a bowl and whisk until light and thick. Whisk in the butter. Mix together the yoghurt and bicarbonate of soda and stir in to the mixture immedi-ately. Whisk the egg whites until stiff and then fold them lightly into the mixture. Then add the sifted flour and baking powder, and fold in. Pour into the baking tin and smooth the top. Bake at 190°C/375°F for 25 minutes, or until the cake springs back when pressed with a finger. Leave to cool in the tin.

Place the honey, water, lemon rind and cinnamon stick in a saucepan and heat gently for a few minutes. Strain the syrup over the cake and sprinkle with the almonds. Cut into 24 pieces.

DRINKS
Bees Knees
1 part liquid honey
2 parts lemon juice
2 parts orange juice
8 parts gin

Put this little lot into a shaker and shake it around with cracked ice. If you use white rum instead of gin the drink becomes a Honeysuckle; if you use brown rum it's a Honey Bee. Whichever way it's interesting!

Honey Comforter
Juice of 1 lemon strained
3 teaspoons honey
1 dessert spoon whiskey, rum or
    brandy (optional)
2 crushed cloves
1 cinnamon stick
Hot water

A good recipe for anyone suffering from a cold!

Put the first four ingredients into a tall, thin glass – an Irish coffee glass is ideal. With a spoon in the glass, carefully add very hot water to taste. Stir.

Making mead

Mead is one of the oldest drinks known to man. It is also said to be a potent aphrodisiac!

The ancient Greeks held a once or twice yearly orgy named a 'Dionysia' with the help of mead, and closer to home our ancestors often made a point of drinking mead for a fortnight before their honeymoons. On the night in question the bridegroom would be filled full of mead in the hope that he would produce a son – and if he did it was considered a particularly good potion.

However in the old days it was probably a pretty unpleasant concoction due to the wild yeasts that were used to make it ferment. The basic ingredients of fermented honey and water remained the same, but taste would have been pretty inconsistent and in Elizabethan times would have been disgustingly sweet and strongly flavoured by modern standards. The better off classes liked to add all sorts of odd things to what they termed their 'pyments' which played a dual role as medicines and alcoholic drinks.

Mead began to lose its popularity when the New World and slavery made sugar readily available. Up until the 18th century sugar was a luxury enjoyed by the rich while honey was the common people's sweetener. As demand for honey lessened, so did demand for the mead wines and ales (yes, they made beer with honey and hops as well) which were supplanted by grape wines and malt beers.

Mead is about 12–14 per cent alcohol and is recognized as an acquired taste – some people just can't stand the taste or smell which is usually sweet, while others love it. The problem if you make it is that you need patience. Mead should have at least two years to mature; in fact the longer the better. Like fine wine or port, you have to wait for the best things in life.

Mead is basically made from yeast, water and honey, while interesting variations include Melomels (fermented honey and fruit juices) and Metheglin (fermented honey with herbs and spices).

Making Mead Yesterday

'To 13 gallons of water put 30 pounds of honeye. Boyle and scum them well. Take rosemary, thyme, bay leaves and sweet briar – one handful altogether. Boyle an hour, put into a tubbe with little ground malt, then stir till lukewarm. Strayne through a clothe and put into a tubbe agayne; cut toast and spread over with good yeast and put into tubbe also. When liquid is covered with yeast put into a barrel. Take of cloves, nutmegs and mace an ounce and a half; of ginger slice an ounce. Bruise the spice, tye it up in a rag, and hang it in the vessel stoppynge it up for close use.'

Making Mead Today

Making mead today should be less hit and miss, and the results should certainly taste better! You will need a wine or mead yeast, available from home brewing shops; and you must start it fermenting two days before you add it to your unfermented honey and water mixture (the 'must'). You will also need water, and it is strongly recommended that you use clean, filtered rainwater rather than turning on the tap.

Dry Mead

The ingredients are 3lb (1.36kg) crystallized honey (3.5lb [1.58kg] if liquid honey); enough water to make up 1 gallon (4.5 litres); and 0.5oz (14.17gm) of yeast.

Measure and mark the level of a gallon of water in your preserving pan. Tip this water away. Now tip in the honey; add water to dissolve it; and continue adding until the gallon level is reached.

Bring to the boil and simmer for five minutes, no more. Stand the mixture aside to cool a little, and filter it through a jelly bag or something similar into the fermentation vessel.

When lukewarm add the fermenting yeast culture that you have prepared. Plug the neck of the vessel with clean cotton material for three days, and when the first vigorous ferment has died down either insert a glass airlock or cover with three layers of cotton material.

Keep in a warm temperature – 65°–80°F is ideal. When fermentation ceases stand the vessel in a cool place for a month and then syphon off into clean bottles and cork.

Lay the mead down horizontally to preserve the corks. The longer you can keep it the better. Some of the best meads are matured for seven years.

Sweet Mead

The ingredients are 4lb (1.8kg) crystalized honey dissolved in water to make 1 gallon of liquid (4.5lb [2kg] for liquid honey); 0.5oz (14.17gm) yeast.

The method is the same as for dry mead. If when you siphon off at the end of fermentation the mead is not sweet enough, add another 4oz (113gm) of warmed honey and stir up the contents of the vessel. Replace airlock and leave for a further three weeks.

Lemon Mead

The ingredients are 4lb (1.8kg) crystalized honey; 1 gallon (4.5 litre) clean rainwater; 0.5oz (14.17gm) yeast; 1 lemon.

Boil the rainwater for five minutes. Let it cool to around 120°F – too hot to leave your hand in it. Add the honey and stir it well to dissolve it. Add the juice of the lemon.

Add the already fermenting yeast to the lukewarm mixture; then cover and let it ferment for three days. If possible transfer the mixture to a vessel which can be stopped with an airlock; alternatively choose a vessel with as small a neck as possible, fill it almost to the brim, and cover with three layers of

clean cotton material tied firmly down.

Leave to ferment in a warm place. When fermentation ceases, transfer to a cooler place for 1 month before bottling off.

Rosehip Mead

The ingredients are 4lb (1.8kg) crystallized honey; 3lb (1.36kg) rosehips; 2 lemons.

Boil the rosehips in 1 gallon (4.5 litre) of water for five minutes. When cool mash the rosehips with your hands or a wooden spoon, and strain the mixture through two layers of butter muslin.

Make the liquid back up to 1 gallon by adding boiled water if necessary.

Add the honey, stirring well to dissolve it, and the juice of the two lemons.

When lukewarm add your fermenting yeast, and then allow the mixture to ferment, settle, etc as in the previous recipes.

Metheglin

The ingredients are 5lb (2.26kg) crystallized honey; 1 gallon (4.5 litre) water; 1 lemon; 1 sprig of rosemary and 1 sprig of balm; ½oz (14gm) grated root ginger; ¾oz (21gm) yeast.

Simmer the herbs, spices and rind of the lemon in the water for 20 minutes. Strain and pour it onto the honey.

When lukewarm add the juice of the

Fermenting mead with airlock and the ingredients for making mead.

lemon and the fermenting yeast. Cover and leave the mixture to ferment for 24 hours.

Pour into a fermentation jar and insert an air lock. Leave to complete fermentation in a warm place; then remove to a cooler place to settle for three weeks before syphoning off into bottles.

You can vary the herbs and spices to suit your palate. Cloves, oranges (juice and peel), cinnamon, marjoram, balm, rue and hops can all be used in this manner.

Wax extraction

After honey wax is the bees' main by-product – a hive can produce 1lb+ (0.45kg+) of wax for your use each season.

Every time you spin honey off from the frames you are left with the cappings, and these should be stored in a plastic container along with any wax left from filtering the honey.

There is also wax gained from cleaning up the hive in places where the bees have been building out of line and making brace comb, and the whole lot can be saved up and sold to a wholesaler or retailer, usually on a barter basis for goods supplied.

Wax cappings mixed with pollen are also considered an alternative medicine (page 164), the most common of which is a teaspoon taken a day to build up hayfever immunity.

The best answer is to use the wax to make up your own polishes and other domestic recipes to make your own sheets of foundation (most local associations have presses); or to make candles which are dealt with overleaf.

Whenever preparing these mixtures be careful because they are sometimes extremely inflammable, in particular beeswax and turpentine mixtures. Don't heat them over a naked flame, and for preference use a double saucepan or 'bain marie' which is a bowl surrounded by hot water within a saucepan.

A final tip is that as with mead, rainwater is always preferable to tapwater.

Extracting Wax

Wax taken from the hive is full of all sorts of bits and pieces which have to be removed before you have pure wax. The easiest way to do this is with a solar extractor. Patent designs can be bought in bee shops, or you can fairly easily make your own.

A solar extractor is basically an insulated container with a double glazed lid. The old cappings, foundation, scrapings, old frames, etc can be piled in on

top of a metal grid over a dark coloured tray. The extractor is put out on a sunny day facing directly at the sun, and with the lid closed the inside temperature soon soars past the 145°/147°F degree stage at which wax melts.

The melted wax runs down the tray, where it is filtered by a fine mesh and drips into a container (any used plastic food containers are fine) leaving the residue behind. Keep filling the solar extractor with more wax so that the residue does not start to burn in such

A home made wax extractor. With the double-glazed lid closed and the sun out, the heated wax drips through a filter into the plastic carton below.

fierce heat.

When you have finished, the wax block should be stored for future use. Wax can become brittle in extreme cold, so a moderately warm room is best. The cartons must be sealed or the wax put in sealed bags in order to keep the wax moth at bay.

Bees' wax and polish

POLISHES

A layer of beeswax polish gives a mirror finish combined with a heavenly smell. It is the purest form of polish you can use; it does however require extra elbow grease.

Metric conversion:
1oz = 28.35gm
1pt = 0.56 litre

Floor polish

8oz beeswax
1pt turpentine

This is a highly inflammable mixture, and a bain marie type of pan should be used for safety with a solid element (electric cooker or Aga style) rather than naked flame providing the heating.

Gently heat the wax and turpentine, until the former has melted (160°F). Stir the mixture well and allow it to cool. Transfer to a suitable container.

Furniture polish

4oz beeswax
1pt rain water
8fl oz turpentine
1 level tablespoon bicarbonate
1 small piece of soap

Shred the beeswax and soap.

Dissolve the bicarbonate in the water and warm it gently with the beeswax and soap until all has turned to liquid. Pour the mixture into a large basin, adding the turpentine which has been warmed in a bain marie.

Keep stirring until it takes on a creamy consistency, and then pour into suitable containers while still warm. Some colouring can be added if necessary.

This mixture can also be used as a floor polish. By increasing the amount of turpentine to 1pt it becomes workable as a leather polish. Some experimentation with quantities is always worthwhile for different uses.

Dubbin waterproofing

4oz beeswax
2oz mutton fat

Dubbin is a traditional waterproofing for leather boots and shoes. Melt the beeswax and fat together in a bain marie. Stir it well, and then pour into a suitable jar. When you use it, warming it a little will make it easier to work into the leather.

COSMETICS

All kinds of cosmetics can be made with beeswax including lipstick, eye shadow and mascara. However these days few ladies are willing to trust their complexions and appearances to home made gooey concoctions, with the notable exception of moisturizing cold cream.

Cold cream

1oz beeswax
1oz Vaseline
1fl oz rain water
½ teaspoon of borax
5fl oz mineral oil

Melt the beeswax, vaseline and oil in a bain marie. Then dissolve the borax in the water, and stir into the mixture.

You can add a trace of scent when nearly cold. Finally turn it into a suitable container.

Cleansing cream

1.5oz beeswax
6oz paraffin
4oz rain water
½ teaspoon of borax
Rose water

You can also try beeswax cleansing cream. Shred the beeswax and melt together with the paraffin, using a bain marie. Then take if off the heat.

Boil up the water and dissolve the borax into it, and then slowly stir it into the wax and paraffin. Add a few drops of rose water, and pour it into a suitable container.

Bees make wax! This small swarm left in a basket took less than a week to make this much of its own wax comb.

Making candles

Beeswax candles are the finest candles. Not only do they have a heavenly smell, but they also burn a long time. They are however very expensive to buy, which is why the beekeeper with excess wax can do nothing better than make his own.

There are four potential types of home-made beeswax candles – foundation, poured, dipped and moulded. They all need wicks, and cotton braided wicks are readily available which are suitable for the job. However beeswax candles need at least double the wick thickness recommended for ordinary candle (paraffin) wax, since beeswax has a much higher viscosity.

Wicks can be stretched tight or left slack; it is just a matter of experimenting to get the right 'burn'. All the ones commercially available will have been 'pickled', and the most suitable are obviously those which have been specifically pickled for beeswax. Church candle manufacturers may be able to supply these, but otherwise they are unobtainable.

Equipment
Any heating of beeswax is potentially dangerous, for it ignites easily. Work in a draught free area and keep fire extinguishing materials near at hand – a bucket of sand or a fire blanket would be ideal; never use water which will just spread the blazing wax.

The wax has to be melted to be worked, and this should be done in your 'wax pot' (once a wax pot always a wax pot) which is set up as a bain marie in a pan of hot water which is regulated to keep the wax molten. Never use a naked flame.

Beeswax only melts slowly and will discolour if it starts to burn (185°F+) so great care and patience must always be taken. The pot should also be cleaned out after each session; either pour the wax out or wait for it to set and break it out.

You will need a ladle to pour the wax, oven gloves to handle anything that's hot, and a hanging line for your candles if using the dipped or poured methods.

If the wax is spilt, it is most easily removed from surfaces when it is just setting; if it is allowed to go hard you have to chip it off. Warm wax spilt on bare skin may feel pleasant, but molten wax will burn. Cool it by immediately dipping the affected area in cold water (not if it is alight, in which case it should be smothered).

Foundation Candles
These are the simplest home-made candles of all, and they burn well owing to the air trapped inside the roll of foundation. You don't have to collect beeswax to make them – simply use sheets of unwired foundation which are relatively cheap.

Foundation becomes brittle when cold, so it is best if you can work it in a warm room where you will use a clean, flat, warm surface for the rolling.

Cut the wick a little longer than the foundation – lengthways for a tall, thin candle; breadthways for a short, thick candle. If possible dip the wick in molten beeswax before starting. This will make it adhere to the foundation more easily, and it will burn more steadily.

Lay the wick along the sheet about 5mm from the edge which should be folded over to cover it; and then roll the whole sheet around the wick as you would a length of material, taking care to keep the roll tight.

Finally secure the candle by warming the loose edge (hold it near a radiator or similar). You can then smooth it down with your fingers and allow it to adhere.

Rolling lengthways or breadthways will make a straight sided candle. You can also make tapered foundation candles by rolling the foundation diag-

MAKING FOUNDATION CANDLES

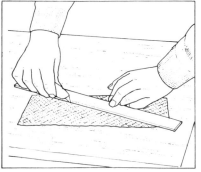

Cut unwired foundation diagonally to make cone-shaped candles.

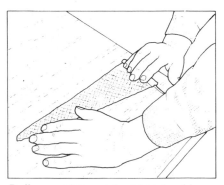

Roll around the wick which should be pickled in wax for best results.

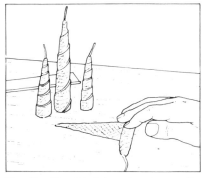

Candles like this burn cleanly, smell beautiful, and last at least an hour.

onally, from corner to corner. Decoration can be added by cutting off thin strips and pieces, and pressing them onto the outside.

Poured Candles

This was the traditional method of making commercial candles which is now considered too tricky and time consuming. For home-made candles it is a method which requires skill. You can use it to make candles as big as you like, but a large quantity of beeswax is needed and there may be a lot of splashing and mess.

The basic technique is that the wick is cut to the length required. It is then dipped into molten wax; cooled; and dipped again until the waxed wick is solid enough to form the basis of the candle. Molten wax is then poured over and adheres to it, until the candle is considered big enough with layer after layer of wax added.

To start, use your bain marie 'wax pot' to melt the wax, heating it to its molten state 160°–170°F. Immerse the wick in it until it is fully impregnated, then draw it out straight and hang it up to cool using a clothes peg and a line. Keep it in the cool away from draughts, for any temperature differences will tend to distort it.

Dip it a second time, but only for a second or two or the original wax will start to melt. Hang it up to cool again.

Next lay the waxed wick on a smooth, cold surface, and roll it with a similar surface (a sheet of glass will suffice) to make it round and smooth. Temperature is critical – if it is too cold the wax will break; if too warm, the wax will come off on the roller. To keep it to the right temperature you can dip it in hot water as the need arises.

The next step is the actual 'pouring'. Hang the waxed wick over your wax pot, spin it round on its wick, and use your ladel to pour molten wax down it. Continue to build up the candle this way, and if it distorts roll out the bumps by rolling and moulding it with the flats of your hands. If it becomes really uneven you can make it workable by submerging it for up to an hour in hot water.

Wax hanging off the bottom of the candle may be cut off and put back in the wax pot for reuse. If the candle gets heavy you should knot the bottom of the wick to prevent the candle slipping down it.

When the candle is the required size, allow it to cool. The top and bottom of the candle can, if required, be moulded to a nice shape by rolling against the hot side of a kettle. Trim the wick to about half an inch, and leave the candle to mature for 24 hours.

The main problem with pouring is that much of the wax can drip and slide off. An easier variation may be to dip your wick as described; suspend it in a mould; and pour in the molten wax around it.

Dipped Candles

For this technique you need a tall, narrow 'dipping pot'. A laboratory measuring cylinder made in heat resistant plastic is perfect for the job.

Melt your wax in your wax pot bain marie, and fill the dipping pot which should also stand in a pan of hot water to keep the wax molten with the water temperature recorded by a thermometer – around 170°F to start with. You will need to keep replenishing the dipping pot as the wax is used up.

Wax your wick by lowering it into the dipping pot, waiting until all bubbles have ceased. Then hang the wick to cool.

Dip the wick again in one fairly quick movement; then roll it using flat, cold surfaces as for poured candles, if necessary dropping it into the hot water for a moment to regulate the temperature.

Dip a few more times with the temperature reduced to around 160°F and the candle will start to build up. Roll after each alternate dip to keep the candle straight until the candle is about 5mm thick; from then on you only need to roll occasionally, using your hands. Cut excess wax off the bottom.

For the final couple of dips turn the heat up a little more to get a smooth outer finish. You can then mould the

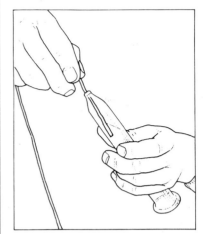
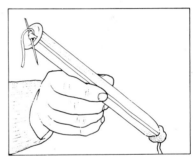
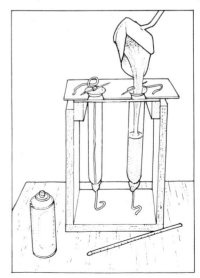

ends and trim the wick as for a poured candle.

Moulded Candles

The easiest and best finished candles are made with moulds which you can buy made in metal, glass, rubber, plastic, etc. This is easy with paraffin candlewax, but moulding with beeswax is quite tricky as it tends to stick to its mould.

The mould should therefore be lubricated with glycerine, liquid detergent or silicone spray sufficient to give complete release, but without overdoing it which might create bubbles in the finished candle.

The mould should be warm when you pour in the molten wax around the wick, heated to around 160°F (20°F past its melting point) which will ensure it contracts well away from the mould when it cools. It may also contract away from the wick, in which case you have to enlarge this central cavity (try pushing down a darning needle) and fill it with more molten wax.

If your mould releases easily, you can pour in the molten wax when it is near to setting (around 155°F) and just about to form a skin on the top. This means there will be little ensuing contraction so the decorative features of the mould are copied perfectly.

The mould should be cooled slowly before release, standing upright in a suitable container wrapped around with newspaper. The wax should be completely cold when it is demoulded, and this may mean leaving it overnight. Beware that even though the outside has set, the inside may still be molten.

Honeybee medicinal products

Propolis is the gooey stuff which the bees use to glue up their hive. Given the time, they will stick their boxes together and the frames to the boxes, and that is why you need the wedge end of a hive tool to open them all up again.

The bees make propolis by collecting sticky resinous globules from the bark and leaf buds of certain trees. The name is derived from Greek – pro meaning 'in front of' and polis meaning 'city'. Quite simply it is a substance which the bees use to seal up and protect their city.

The first thing they do when they occupy a new hive, or any hollow nesting place in the wild, is to coat the inside walls of their home with a very thin layer of propolis, so thin that it cannot be seen with the human eye. They also coat the inside of every wax cell before use for laying or filling with honey in a similar fashion. The reason is thought to be that propolis contains natural antibiotics which protect the colony, staving off brood diseases including bacterial and fungal growths.

They use propolis to glue everything together, and will mix it with wax to fill up any undesirable holes. If an intruder still manages to get in they will hopefully kill it, and if it's too large for them to push back out (such as a mouse) they will mummify it in a resin and wax casing so they don't have to cope with a rotting carcase.

Removing Propolis

When a hive is well propolised, take great care not to do any damage when you come to opening it up. The hive tool should be inserted carefully to avoid splintering the wood, and each box must be twisted before you attempt to lift it clear; otherwise you will lift it with half the frames from the box below still propolised to it.

Propolis in the boxes and on the frames can easily be removed by scraping with the sharp end of your hive tool. The metal ends on frames are always heavily propolised, and are most easily cleaned by being dropped in boiling water with some washing soda added. Plastic bee escapes should be soaked in meths to clean the propolis off.

Propolis as an alternative medicine

Propolis is known to have healing powers, though as with most alternative medicines its claimed powers are almost certainly greater than its true powers.

Some beekeepers collect all the propolis they have scraped off, and eat a small amount as a matter of course every morning. A popular method is to scatter it over breakfast cereal; it is extremely unlikely to do you any harm, and it is quite likely it does you good.

In the past propolis was widely used as a surgical antiseptic, from very early times up until the Boer War. It lost its popularity with the advent of modern medicine, and has now returned to favour with many other alternative medicines. In particular it is highly regarded in eastern European communist bloc countries which have led modern propolis research, but the initiative in marketing has predictably come from the capitalist 'west'.

'Comvita' is a major New Zealand producer with an attractively packaged range of propolis based products that include Propolis Herbal Elixir, Propolis Tincture, Propolis Ointment, Propolis Toothpaste (!), and Propolis capsules.

These products are described as 'all natural infection fighters' which are variously said to help throat infections and the common cold, mouth and gum disorders, gum decay (if you use propolis toothpaste), some skin conditions, resistance to general illness, pain from stomach ulcers, minor infections, and fungal skin complaints.

Propolis has also had another unusual historical use. The great violin maker Stradivarius favoured it to varnish his now almost priceless and perfectly preserved violins!

Other medicinal products

The pollen collected by bees is also known as a valuable tonic supplement containing proteins, carbohydrates, vitamins, minerals and enzymes. It is widely available in health food shops,

usually packaged as dried pollen granules.

Pollen is the living male sex cell of plants and trees which allows the bees to pollinate as they forage (page 75). In the hive it is used to make brood food to feed the larvae and young bees, and it is estimated that a normal size colony could consume 40lb (18.14kg) of pollen a year.

Despite this, it is possible to collect pollen by means of a 'pollen trap' at the entrance which scrapes it off their legs as they go in. There is usually surplus, but at times of pollen shortage some will have to be fed back to the bees and is then usually mixed with an artificial pollen. This is rarely necessary in the UK.

When taken by humans, pollen collected from hives is variously claimed to act as a stimulant for those recovering from illness or surgery; to help impotence and alcoholism; and as a help for anaemia, colitis and gastritis. On a more prosaic level it is renowned as a protector against hay fever;

however some experts say this only works if you take it in your 'own back garden' where the bees have collected it from the sources that are causing you all your trouble.

Royal Jelly

You are also likely to find royal jelly products in your local health shop. This specialist brood food is made by the bees and is used to feed all new larvae for the first two or three days. From then on it is reserved exclusively for future queens, while workers are brought up on more prosaic pollen based brood food.

Royal jelly is a milky liquid which can be collected by cutting off queen cells, removing the larvae, and scooping out the contents which can be stored long term in the fridge. Specialist techniques need to be used to produce large numbers of excess queen cells to do this, and commercial royal jelly production relies on queen rearing on a massive scale. Approximate productivity is likely to be 1oz (28.35 gms)

Some Comvita honeybee health products. Honey & Boysenberry Conserve (for a sweet tooth); Propolis Lozenges (for the throat); Propolis Tincture (for gargling or cuts); Pollen Granules (shake on your cereal); 100% Propolis (take regularly to improve immunity); Propolis Ointment (all purpose skin application); Propolis Elixir (herbal cough syrup); Beenut Butter!

of royal jelly collected from 100 or more queen cells – no wonder it's so expensive!

Royal jelly is a claimed 'cure-all' which is normally mixed with honey and sometimes propolis as well. It is said to improve body resistance against illness or infection, and some believe that it stops you growing older! Like other bee products it is safe to say that this splendid sounding product does you no harm and probably does you some good; though whether it is worth the high price is a matter of opinion.

The Honeybee's Problems

Earlier parts of this book have touched on the predators and diseases which could afflict your colony. Hopefully and probably this will never happen, but if it does you will need to know signs, symptoms and cures. This section identifies the major bee colony afflictions such as acarine mite, wax moth, nosema, varroa and foul brood, and tells you what you can do about them. It must be stressed that if you have any doubts you must consult expert opinion, for it is the duty of every beekeeper to contain the spread of these problems.

However we start overleaf with a problem of a different nature – the 'Killer bee'. This unwelcome type of honeybee has not been seen in Europe, but is presently threatening a large part of the American continent. . .

The simplest guide to healthy beekeeping is to watch your colonies and make a proper inspection of your brood frames on a regular annual basis. If anything is amiss you should soon pick it up, and hopefully forestall it before the problem spreads to other colonies. If you suspect something but don't have the knowledge to be sure, call in a second opinion or send suitable samples away for analysis. Hopefully on inspecting the brood you will see nice clean, healthy frames like the one shown – a little honey in the corners, the queen conveniently marked with a blob of yellow, and all's well!

Africanised 'Killer Bees'

South American beekeeping has been thrown into confusion by the arrival of Africanised 'killer bees'. These bees are the unfortunate result of genetic experiments with African bees which had been imported for use at a Brazilian research centre in 1956.

In 1957 26 colonies escaped, composed of 'Africanised' bees interbred with Apis Mellifera which was the dominant honeybee throughout America as it is throughout Europe. Though appearing virtually identical (experts can tell them apart by the size of the forewings, pattern of veins in the wings, and the wax glands in the underbelly) the Africanised versions inherited the worst traits of the super tough Central African sub-Sahara honeybee which has evolved into an aggressive survivor used to existing in difficult conditions.

The result has been that Africanised bees have resolutely spread through South America, and by 1987 had reached southern Mexico and were threatening the southern states of the USA. Having escaped from Sao Paulo in 1957, those 26 colonies have grown and grown, mating with and taking over from Apis Mellifera which they rapidly outnumber. They breed faster, fly faster, nest anywhere, and in Charles Darwin's terms have simply taken over by natural selection.

The problem is that many of those South American countries have depended on honey production for their income. Venezuela and Guatemala were turned from honey exporters to honey importers within a couple of years of Africanised bees arriving, and the economy of Mexico as the world's number one honey producer (worth $50 million a year) was under threat at the time of writing. The Africanised bees effectively destroy the colonies of Apis Mellifera and take all the honey, and South American beekeepers have proved totally unable to 'work' them with primitive protective gear.

'Killer' bees

The major problem in attempting to work them is of course their well publicized killer instinct.

The African traits of Africanised bees make them excessively aggressive defenders of their colonies. Normal honeybees will defend a few metres surrounding a colony, and only do so when severely provoked; Africanised bees will defend a 200–300 metre area, and are much quicker to attack if they

African honeybees were imported to a Brazilian research station in 1956. In 1957 26 interbred colonies escaped from Sao Paulo. They rapidly spread north and south. In the north they had reached Tehuanetepec in 1987 and were expected to reach the southern tip of Texas by 1990. In the south the progress slowed as the weather became colder; having originated from the sub Sahara they flourish in hot conditions.

feel threatened. Their sting is no more lethal, but they will sting and continue to sting in much greater numbers, carrying on an attack for up to 30 minutes.

Humans and animals can only tolerate so many stings, and it is this which has led to fatal results. Statistics are extremely vague, but it is possible that up to 40 people have been killed in this way since the bees first escaped.

Solutions

Africanised bees are genetically most suited to hot climates. Research has shown that the further they get from the Equator, the less African they become. So far they have shown no inclination to colonise colder areas beyond 34 degrees south of the Equator down into Argentina, and experts believe they will also show no inclination to go beyond 32 degrees north.

This should mean that the northern states of the USA are all safe, as is northern Europe so long as the frightening possibility of the Africanised bee genetically adapting to colder climates is never realized. As for their southern states, the USA started fighting a rearguard action with a three year $8.6 million aid programme for a Bee Regulation Zone across the narrowest point of Mexico which was started in 1987. Their idea was to use bait hives to capture and kill all northbound bees, but while a sound plan in practice many believed this could at best only slow Africanised progress for a couple of years. Many swarms would be missed since they have a tendency to nest anywhere, and to be effective the aid programme would need to last indefinitely.

In Mexico the 47,000 or so families who make their living from bees are being educated to cope with the Africanised version. Having worn little protective clothing in the past, those under threat are now encouraged to dress fully with leather boots in order to cope with any attacks, but with widespread illiteracy and poverty it is by no means simple for the authorities.

Their neighbours in the southern USA should have considerably more resources to cope with the problem. They believe that Africanised bees can be used for honey production, but only with much more sophisticated handling which would spell the end of beekeeping as a cottage or hobbyist industry in those areas. Even then they would be competing with wild swarms which expand to use up all available food, and it is feared that a likely anti-beekeeping backlash could make the invasion of the Africanised bees even easier by removing any bee competition.

Until a way is found to eliminate the results of that unfortunate genetic experiment, people are just going to have to find a way to live with them in the areas in which they flourish.

Defensive behaviour of the Africanised honeybees during honey removal, Boaco, Nicaragua 1988.

Brood diseases

Hopefully your hive will never experience the kind of problems which are summarised here. They are mostly unusual, but it is as well to have some idea how to recognise them if you feel you do have a problem with your bees.

Healthy brood is easily recognisable. The larvae are at first shiny and white, lying coiled in a C shape in their uncapped cells. At about six days old the cells are capped over by the worker bees to allow the larvae to continue developing. The cappings vary from light to dark brown and are slightly convex.

If in doubt, frames can be submitted with a separate covering letter to the National Beekeeping Unit (Luddington Experimental Horticultural Station, Stratford upon Avon, Warwickshire CV37 7SJ) for professional examination. Remove a suspect brood frame, shake the bees off into the hive, and pack it securely in plenty of newspaper with metal ends removed in a strong cardboard box. Do not use polythene, plastic or a tin box which may cause the comb to sweat and decompose too rapidly for any worthwhile examination to be done.

ADDLED BROOD

Addled brood is caused by a genetic fault in the queen. It kills advanced pupae and bees about to emerge from their cells. The diseased bees are recongisable by having very small abdominal parts, while their cells may have dark moist, sunken or perforated cappings.

The remedy is to requeen the colony with an unrelated queen.

BALD BROOD

This is brood with the cappings removed as a result of the activities of wax moth larvae who start to chew away the wax; the worker bees then remove the rest of it. Some of the new born bees may have crippled wings or legs, but otherwise there is only slight overall damage. Wax moths are dealt with in

more detail on page 175.

Sometimes extensive bald brood can occur with no wax moth larvae present, in which case the colony should be requeened with an unrelated queen.

CHALK BROOD

This is caused by the combination of a fungus and slightly chilled brood, and takes its name from the chalk white appearance of the dead brood.

The larvae eat the fungus with their food and subsequently die. At first they puff up; later they shrink and become hard. They usually stay chalky white, but can turn dark grey. The disease can be recognised when the cappings have been removed by worker bees, and is usually intermittent throughout the brood.

Careful manipulation of the colony to guard against chilling is usually an effective way to contain the disease, and it should clear up during the course of the season. Badly affected frames are best destroyed. If the disease persists the colony should be requeened with an unrelated queen.

CHILLED BROOD

Chilled brood can occur if there is a sudden spell of cold weather in spring. The bees retreat towards the centre of the brood, leaving the extremities of the frames unprotected. The beekeeper can create this problem himself by examining frames in too cold weather, or by excessive sugar syrup feeding early in the season.

The result is that unprotected larvae of all ages die. The younger dead larvae often become shiny black and can be seen lying in the bottom of their cells; older larvae and pupae become greyish under moist, sunken cappings.

The worker bees can usually clean out the dead brood, but if the problem is extensive the frames should be removed, replaced by other frames, and stored in a dry area until the dead brood becomes hard and shrivelled. The

frames can then be returned to the hive for final cleaning by the bees.

FOUL BROOD

Foul brood is the most serious brood disease, and can be American or European foul brood. The labels 'American' and 'European' don't mean that one strain occurs in America and the other in the USA; they can both occur anywhere.

The former is a disease of sealed brood and is by far the most dangerous brood disease in the UK; the latter is a disease of unsealed brood, which can also spread rapidly and be difficult to eradicate. Both require vigilance and immediate action.

American Foul Brood

AFB is caused by a spore forming bacterium Bacillus Larvae. The spores are ingested by very young larvae where they germinate rapidly. The larvae then die, but millions of AFB spores remain. These spores are very resistant to heat, cold and disinfectants; and they retain their power of germination for many years in honey, old combs, old hives, etc.

Infected larvae die within their sealed cells. The cappings become sunken and are perforated by the adult bees; some may be moist and very dark. At this stage the remains of the larvae are slimy brown. If you push in a matchstick, they can be drawn out like a mucus thread. This is one of the tests of AFB.

The remains will gradually dry up in their cells, becoming darker in colour. Eventually all that is left is a very dark brown, rough scale laying on the lower side of the cell. These scales can be seen by holding a diseased frame against the light. The worker bees find them very difficult to remove, and they remain for many years in frames from colonies which have died out as a result of AFB.

If any contaminated frames are transferred to another hive, that colony will become infected. The diseases can also be spread by the beekeeper and his equipment; by robber bees from diseased colonies; by drifting bees; and by infected swarms.

The best control is regular examination of the brood, and by the burning of frames from infected colonies with all the bees destroyed. The hives and their components must be sterilized with a blowlamp, as must all the beekeeper's hardwear. Clothing, etc must be thoroughly washed in hot, soapy water.

European Foul Brood

EFB is caused by the bacterium Melissococcus pluton which multiplies in the mid gut of larvae.

The larvae die when about four days old, with their cells still unsealed. They can be found in unnatural positions in the cells, looking yellowish brown at first and eventually becoming a loosely attached brown scale. The disease cannot however be positively identified visually and is easily confused with other less serious brood problems. The gut contents must be microscopically examined to confirm the presence of EFB.

EFB is spread in the same way as AFB by both beekeeper and bees. Colonies with a high proportion of dead brood should be destroyed; colonies which are only lightly infected may be treated with an antibiotic, but by law in the UK this must be administered by an Appointed Officer after approved examination.

Controlling Foul Brood

AFB and EFB come under the Bee Diseases Control Order of 1982. In essence this states that any beekeeper who suspects the presence of AFB or EFB must contact the local office of the relevant Agricultural Department to have the colony examined on site. In the meantime do not open your hives.

A sensible precaution is to insure your colonies against complete loss through foul brood. In the UK the BBKA and local beekeeper associations have low cost schemes which are often included in the cost of membership.

Below left: American Foul Brood. 'Scales' of dead larvae; oblique view of comb, as seen from above, showing scales in lower angles of cells.
Below right: European Foul Brood. Comb with many dead larvae.
Bottom left: Chalk Brood. Comb with dead larvae in uncapped cells.
Bottom right: Drone brood in worker comb, from a drone-laying queen.

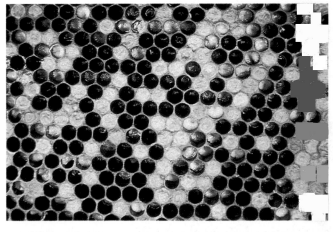

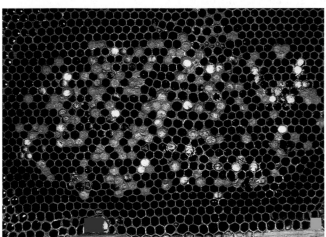

However prevention is the best policy of all, and the Ministry of Agriculture, Fisheries and Food list 10 guidelines to control the spread of foul brood which should be adhered to by all beekeeprs:

1 Keep the apiary clean and tidy.
2 Never throw propolis or brace comb on the ground. Place it in a container, and remove it.
3 Never buy secondhand frames.
4 Never buy colonies unless you know they come from disease free apiaries. Never accept stray swarms of unknown origin.
5 Always sterilize secondhand hives.
6 Never feed honey from doubtful sources or allow bees to gain access to it.
7 If the colony dies during the winter and the trouble is not due to starvation, close the hive pending examination of a sample frame so that remaining stores cannot be robbed.
8 Never exchange frames from one colony to another unless you know all your colonies are free from disease. Where possible supers should be marked and always used on the same colonies.
9 Take care to prevent robbing at all times.
10 Arrange the hives to reduce drifting to a minimum.

NEGLECTED DRONE BROOD

A drone laying queen or drone laying workers (page 134) will produce irregular patches of drone brood in worker cells. The colony is likely to be weak; the drone larvae will be under-nourished and will eventually be neglected so that they die and decompose.

The frames often have an untidy appearance due to the irregular laying, and the cappings are sometimes partially torn away. The decomposing larvae eventually dry up and become dark masses sticking to the sides of the cells.

The remedy is to requeen the colony or unite it with another colony. Badly distorted combs are best destroyed.

SACBROOD

This is caused by a virus which causes the larvae to die in their cells. Their outer skin becomes loose and contains a watery fluid. Later they dry and shrivel like a Chinese slipper.

Sacbrood is noticeable from May onwards when the cappings have been removed by worker bees. No specific treatment is known, but the disease usually disappears towards the end of the season. Only a small proportion of the brood is usually infected; however if the desease is more extensive or persists, requeening the colony with an unrelated queen is recommended.

STARVED BROOD

Colonies which have produced a large amount of brood early in the season can run short of food during bad weather in May or June. Some of the brood may be abandoned, egg laying ceases, young larvae may be eaten by the bees, and pupae thrown out of the hive.

The remedy is to give them an emergency feed with sugar syrup.

STONE BROOD

This is extremely rare, and is caused by a mould that attacks the larvae. It turns them into hard, brown objects lying in open cells. No form of treatment is known, and an infected colony should be burnt.

Other pests and diseases

As with brood diseases, if you have any doubts about the health of your colony consult an expert. Frames, dead bees, and debris scraped from the floor of your hive can all be sent to the National Beekeeping Unit for analysis. What you send depends on where the disease is most likely to be found.

ACARINE MITE

Acarine mites infest the breathing tubes of adult honey bees. The effects are not noticeable during the summer, but in winter and early spring an acarine infested colony is less likely to survive.

The Acarine mite is widespread throughout colonies in the UK, but normally it is only a minor infestation which the bees suppress naturally. The mite flourishes in a poor summer season when the bees spend a lot of time in the hive, and during periods of queenlessness when the worker bees live longer. It is also more likely to flourish in weak colonies, where treatment should be directed at both colony and mite if you are to be successful.

Infested bees can be seen clinging to the stems of plants, or crawling about with fluttering wings. Their abdomens may be distended, and their wings may have a dislocated K shaped appearance. Bees exhibiting these symptoms may emerge from the hive in large numbers on bright days in spring and autumn, but the symptoms may be due to other causes. Microscopic examination of dissected bees is needed to confirm the presence of the mites.

Acarine can be introduced into a colony by drifting of infested workers, by hiving infested swarms, by the beekeeper inadvertently moving bees from an infested colony into a healthy one, etc.

Bees which escape infestation during the first week of adult life are immune since a hard fringe of hair grows up which prevents the mites entering their thoracic spiracle. The spread of Acarine therefore depends on young bees being in close contact with older, infested bees.

In the winter the mites will continue to breed within infested bees, but cannot spread as there are no young bees to spread to. Many of the old bees will die as a matter of natural course, but those that don't will pass on the mites once the queen starts laying. The young bees will then continue to pass on the mites as new bees are born.

During the high season the worker's life is so short that Acarine may have little effect on performance. However the winter bees that are left in the autumn have their usually long lives severely shortened by activities of the mites building up within them, and this can result in severe or total depletion of the colony. If the bees die out totally the mites will as well, for they cannot live for long without their hosts.

Acarine treatment
Treatment should be administered before the colony gets to the stage where it is irreversibly sliding into extinction. Even if you wipe out the mites, the colony may be too weak to produce the necessary new young bees to survive. Furthermore it may be necessary to requeen if the queen appears to be performing unsatisfactorily due to being damaged by the mites.

The easiest treatment is with a smoke strip bought for the purpose from a bee dealer. Two applications are necessary one week apart. One strip treats one brood chamber. They should be given in warm weather in early summer, prior to the main honey flow.

The treatment is ineffective in colder weather because the bees cluster too tightly. It is given in the evening when the bees have ceased flying. An empty super covered by a crown board is placed on the hive over the exposed frames of the brood, and the lighted strip is dangled down through the central feed hole which is then closed over with a suitable block. The second bee escape must obviously be closed, as must the main entrance to the hive. Alternatively you can pin a sheet of newspaper over the hive entrance – the bees will chew through it by the next morning.

If the hive has honey supers on, a couple of frames should be taken out with bees removed to leave space for the smoke strip to hang down. The smoke should be able to circulate quickly through the bees in the hive. It is lethal to the mites; harmless to the bees, brood, and stores. After an hour the entrance can be opened, and the smoke strip super removed the next day.

The alternative 'Frow treatment' is only recommended for use in late autumn or early spring when the bees have been flying but there is cold weather to come. It is much more complicated and has many disadvantages, not least that it is likely to result in the death of many of the infested bees.

Bottles of Frow treatment mixture are available from dealers; it is a poisonous and highly inflammable mixture of petrol, safrol, and nitrobenzene which must be handled carefully.

2 ml of the mixture is poured onto a flannel pad over the central bee escape in the crown board, and then covered with some kind of lid. This must be repeated every other day until the hive has had seven doses. The pad is then left for a further three days.

DYSENTERY
Dysentery means soiling of the hive contents by the bees. While not infectious, it can aggravate other infections such as Nosema and thereby become a serious problem. The bees will also have to clear up their own mess which may interfere with their normal quiet winter period. At worst this may pass the dysentery round the whole colony in a vicious circle, in which case the bees will perish.

Bees can normally stay a long time in the hive without taking cleansing flights. However if there is abnormal faecal build-up due to the presence of too much water in their food, the bees may start to foul the frames when the weather will not permit cleansing flights. Dysentery is therefore mainly a winter problem which can be put right with a timely cleansing flight, but can also occur in the summer when sudden bad weather confines a large number of bees.

The water may be present in unripe honey which the bees consume due to circumstances; in honey which has fermented; or in unripe stores due to the beekeeper feeding them sugar syrup too late in the autumn (first week in October at latest). Note that only refined beet and refined cane sugar are suitable for feeding bees. Brown sugar, raw sugar, etc cause excess water and dysentery.

Dysentery may also be caused by granulated honey which the beekeeper feeds back to the bees – the bees are only able to consume the watery parts of the honey. A late crop of heather honey which does not ripen fully may also be the cause.

Preventing dysentery
Dysentery is easier to prevent than cure. The bees must have suitable stores with a limited water content as described above, and hives must remain free from moisture. They must be rain proof, and sited in a reasonably dry spot with sufficient air movement. In winter the hive must be ventilated with damp warm air constantly being replaced by cool, dry air from below. Methods of preparing a hive for winter are described on page 88.

The only effective cure is a cleansing flight. In mild cases the bees can be fed thick, warm syrup to help them survive until the arrival of suitable weather for flying.

NOSEMA
Nosema Apis is a parasite which lives in the mid gut of the adult bee where it multiplies until the resulting spores are excreted by the bee. These spores can then lay dormant for many months in dried spots of excreta on frames, until they are inadvertently swallowed by another bee which starts the cycle rolling again.

Nosema will spread most rapidly at the end of winter. At this time there is inevitably some soiling of the frames, and healthy workers pick up the infection as they go about their business of doing their own spring clean. The infection will spread quickly; the workers will be less able to perform their tasks; and the colony will not be able to expand as it should. Workers infected by nosema will not live as long as healthy bees, thereby reducing the brood rearing and honey producing capacity of the colony. If the queen becomes infected the colony may becomes queenless or supercede itself. Drones can also become infected, but this is not a serious problem.

Occasionally nosema will wipe out a colony; usually it will survive, with the number of infected bees declining rapidly as soon as regular cleansing flights away from the hive become possible. From this point onwards, the old infected bees will die off and be replaced by healthy bees.

However enough spores are likely to remain on the frames to infect a few bees in the winter cluster, leading to the start of another cycle of nosema. These frames are therefore the most potent carriers of the infection. Infection can also be spread from hive to hive by the robbing of infected hives; or by the beekeeper transferring infected bees or frames to another hive.

Diagnosis and cure

Nosema can only be diagnosed by microscopic examination. A sample should consist of at least 30 bees in a matchbox taken from the hive entrance or crown board feeding hole. Telltale signs such as inability to fly, excreting on the frames, or a heap of dead bees in front of the hive after a cleansing flight may be associated with other problems such as starvation, fermenting stores, or a sudden drop in the air temperature.

Many beekeepers feed the antibiotic drug Fumidil B to guard against nosema when they fast feed in the autumn (page 86), but this will have no effect if nosema reaches the spore stage.

Frames infected with nosema spores must be fumigated with 80% acetic acid which is made by mixing one part of water with four of glacial acetic acid available from chemists. Any acid splashes on the skin must be washed off immediately, and great care must be taken not to splash any into the eyes.

The fumes will not harm honey or pollen. Metal ends will corrode badly and should be removed. Scald them in hot water with washing soda to sterilise them. Put the frames in a suitable brood box or super with the usual spacing. Stack the boxes with square cotton wool wads soaked in 100 ml of the acid laid in the bee space between each set of frames, with a final soaked wad on the top.

Block the entrance, cover the roof, and seal off any gaps through which the fumes may escape. Leave the stack for a week which is long enough for the fumes to kill the spores in even the coldest conditions; they will do it more rapidly if the weather is warm. Afterwards the frames should be well aired before they are ready for use.

VARROA

At the time of writing no cases of varroa had been found in the UK. However having become endemic in Holland, France and Germany, experts forecast that sooner or later the varroa mite would gain a foothold in the UK despite the complete ban on colonies and on imported queens from virtually all countries except New Zealand. Bee colonies are occasionally found on ships; the Channel tunnel would make entry easy; or varroa could be introduced via an illegally imported queen.

Varroasis is the infestation of larvae and adult bees by the parsitic mite Varroa Jacobsoni. The mite spread from the Eastern honeybee, Apis Cerana, to the Western honeybee, Apis Mellifera when beekeepers took AM colonies to the Far East many years ago. More recently the mite has spread through many countries via the introduction of infested bees, reaching every continent bar Australia by 1988.

Varroa mainly infests honeybee larvae. The bees are usually born unscathed, only dying or emerging deformed if the infestation is very heavy. Mature female mites which look like tiny crabs to the naked eye escape with the emerging bees, using adult bees as a means of transport. They live for about two months in the summer, and up to eight months in a winter cluster. The females can clearly be seen on bees and unsealed larvae in heavily infected colonies, but should not be confused with the wingless fly Braula Coeca which has six, rather than eight, legs and is harmless.

Varroa mites live compatibly with Apis Cerana, but infestation can result in the extinction of Apis Mellifera colonies for reasons which are by no means certain. The actual effects of varroa had not been clearly deocumented at the time of writing, though there were some reports of large numbers of colonies becoming infested and dying out.

Detecting varroa

The MAFF has recommended that all UK beekeepers should use the following 'tobacco smoke technique' to check for varroa each October (or failing that in March) when there is little or no brood, ensuring that any mites are exposed rather than shielded in sealed brood cells. It should not be used during a honey flow when it would taint the honey.

3 grams of pipe tobacco (roughly a heaped dessert spoon) is sufficient for a normal sized colony. Overdosing may kill bees.

MAFF Tobacco Smoke Technique

1 Wait until the bees have stopped flying in the evening.

2 Hive floor boards must be free of wax, propolis and debris.

3 Remove the entrance block.

4 A flat sheet of light-coloured paper ('the insert') cut to a size to cover the floor area should be slid into the hive entrance. Decorator's lining paper is suitable, but any reasonably robust paper or soft corrugated cardboard may be used. All types of single-walled hives can be treated in the same way. Some simple modification to the procedure will be necessary where double-walled hives are used.

5 Close the hive entrance with newspaper.

6 Push a lightly complacted piece of newspaper about 400 x 300 mm (16 x 12 inches) into the bottom of the smoker and get it well lit. Wait until the

paper has almost burned and compress it lightly with a hive tool to ensure the smouldering remains cover the mesh bottom of the smoker.

7 Quickly add the pipe tobacco, close the smoker lid and gently work the bellows until tobacco smoke is produced from the smoker nozzle.

8 Make a small gap at one side of the hive entrance and insert the smoker nozzle. Then introduce the smoke smoothly with slow, regular and full squeezes of the bellows. In this way cool smoke is produced and all the tobacco burns in two to three minutes.

9 Close the hive entrance with the newspaper.

10 Early the next morning, before the bees are ready to fly, take out the newspaper blocking the entrance, remove the insert and replace the entrance block.

11 Debris on the inserts of all the hives in the apiary should be bulked together in a suitable container marked with the apiary's identification. Suitable containers are made of firm cardboard. Hive debris should not be packed in tin boxes, plastic bags or other airtight containers, in which decomposition rapidly takes place, or in unprotected envelopes which allow dead bees to receive the full force of the Post Office franking or transit procedures.

12 The packed hive debris should be sent for examination to The National Beekeeping Unit, Luddington EHS STRATFORD-UPON-AVON, Warwickshire CV37 9SJ.

Powers for the control of varroasis are included in the Bee Disease Control Order 1982. When the first outbreak is confirmed a *temporary standstill order* will be applied to apiaries within about 5 km of the infestation for as long as it takes to establish, by inspection, the extent of the outbreak. If the infestation is deemed to be 'isolated' to a very recent importation of bees, then *all* the colonies in the infected apiary will be destroyed. The destruction procedure will be the same as that adopted for colonies infected with American foul brood, followed by the lifting of the standstill order.

However, if other outbreaks are discovered, colonies will not be destroyed and the standstill order will be lifted. The whole country is then considered to be infested, and the Ministry will advise beekeepers how to control the disease.

WAX MOTHS

The method for guarding against possible wax moth infestation is summarised on page 89. The wax should be protected in this manner each autumn which will prevent the greater or lesser wax moth laying and gaining a foothold. As well as using PDP crystals, the frames should if possible be loosely stored in an uninsulated building to take maximum advantage of the cold weather of winter.

If wax moth does become established it will wreck the wax in your frames and

American Foul Brood. Probing a cell to demonstrate the 'ropiness' of a dead larva.

make any honey unusable. Wax moth cannot withstand sudden cold, and the most effective method of killing it off is to transfer the bees and place the frames in a very cold environment. Either put them in a deep freeze overnight, or leave them for a longer period (10 days at 2°C) in the fridge. The frames should be stacked to allow the cold air to circulate freely.

There are other methods using heat, chemical fumigants, and biological bacillus which are considerably more complicated and specialised. The latter does however allow the frames to be treated with the bees in situ.

Beekeeping in history

It seems that man has taken honey from bees since he first set foot on Earth. The oldest records of this are in prehistoric cave paintings, a notable number of which are found throughout Spain. One at Cuevas de la Arana in Valencia shows a man climbing up a cliff to rob a swarm of wild bees. It is estimated that it is around 15,000 years old.

Honey became an important commodity throughout the ancient civilised world – Assyria, Babylon, Egypt, India, Persia, Greece and Rome all made extensive reference to honey in their art and literature. Egypt in particular has a wealth of hieroglyphic carvings and wall paintings which show how highly honey was rated. Few of the great pharaoh's funeral vaults don't feature some aspect of bees or honey, with wall paintings showing details of collecting honey, filling and sealing the jars, or baking honey cakes. Both honey and cakes were left with the sarcophagi as food for the dead.

In Greece nectar was the drink of the gods, and there is extensive reference to bees and honey in Greek literature. One of the earliest is Hesiod, writing around 700BC:

'Every day the bees work eagerly all through the day till sundown and set the white combs, while the drones stay within the roofed hives and gather into their bellies the toil of others.'

Reference to their hives is made by Virgil in Georgics IV:

'As for the hives themselves, whether you have them made of hollow pieces of bark sewn together or woven of pliant withies, let them have narrow entries for the winter congeals the honey and heat makes it liquid. Both forces are equally to be feared by the bees. You must make their homes snug, and fill up their crevices by smearing them all round with smooth mud.'

The elder Pliny (AD23-79) had this to say about the hives of the time:

'The best hive is made of bark, the second best of the fennel plant, and the third of withies. Many have also made hives out of transparent stone [thought to be mica] in order to look at the bees working inside.'

Medieval Europe and beyond

'In this hive we are all alive,
Good liquor makes us funny!
If you be dry, step in and try
The value of our honey.'
      (An old English inn sign)

Long before the Roman invasion, Britain was described by the early Druid bards as 'The Isle of Honey'. Tributes were paid with mead and honey using a quaint measuring system: a milch-cow measure of honey could be lifted up to your knees; a large heifer to your waist; a small heifer to your shoulders; and a dairt over your head.

Honey was much used for cooking and baking in those pre-sugar days, and most importantly for making alcoholic ale or mead drinks (page 158). These held sway over the drinking population for hundreds of years until they were supplanted by hop beers, introduced by Flemish immigrants and variously described as a 'wicked weed which spoilt the taste of the drink and endangered the lives of the people'. Pepys referred to metheglin (spiced or herbal mead) twice in his famous diary – on 29th February 1659 *'A brave cup of Metheglin, the first I ever drank'*; on 25th July 1666 *'I had Metheglin at the King's own drinking, which did please me mightily'*.

Ancient Germany was another major honey country, and upon the introduction of Christianity honey production went into overtime there in order to keep up with the demand for church candles! Taxes were demanded in honey and wax, with the annual honey market in Breslau famous for centuries. There were mead breweries in Munich, Ulm, Danzig and Riga, but honey production went into decline during the period of the 30 Years' War in the early 17th century.

Their next door neighbours in France esteemed honey equally highly, but those in charge seem to have taxed the poor beekeepers to the hilt throughout history! There was a medieval beehive tax as well as a honey hunting tax, and in addition vassals had to hand over a certain amount of honey and wax to their Barons each year.

The situation escalated to the point where French beekeepers began to destroy their hives rather than pay higher taxes which they thought were on the way in 1791. This happened when the government demanded an exact record of hives in every area from their local prefects. The French government introduced a hive tax as recently as 1934 which was designed to assess beekeeping along with farming.

Elsewhere in Europe honey collecting and beekeeping were major industries from the west to the east. Poland was particularly rich in honey. Writing in the 11th century, Gallus remarked:

'*Pane et carne et melle satis est copiosa . . . ubi aer salubris, ager fertilis, sila melliflua.*' (There is plenty of bread, meat and honey . . . where the air is salubrious, the fields are fertile, and the forests flow with honey.)

A little later Holinshed's Chronicles of 1577 claimed that the honeycombs of Poland were so great and abundant that huge bears could sometimes drown in them!

America

'*Wheresoe'er they move, before them Swarms the stinging fly, the Ahomo, Swarms the Bee, the honey-maker; Wheresoe'er they tread, beneath them Springs a flower unknown among us, Springs the White Man's Foot in blossom.*'

(Hiawatha – the 'white man's foot' was white clover)

In ancient times it appears the honeybee Apis Mellifera was non existent in America and Australia. The Americas had smaller stingless bees such as Apis Trigonae and Meliponae, and the Indians cultivated them to produce a thinner style of honey.

Honey taxes were once again prolific! Montezuma, the Aztec ruler of Mexico, was paid a tribute of 700 honey jars which would of course have been large pottery jars. The Mexican mead was named acan, but we don't know if he demanded that from his subjects as well.

Apis Mellifera honeybees were brought from the Old World to the New by Spanish, Dutch and English settlers at the end of the 16th century. In some areas this was a mistake, for they unsurprisingly robbed the sugar cane from the plantations in places like Cuba and Barbados. The planters did their best to annihilate them, and meanwhile they spread across the United States.

They swarmed from the Atlantic to the Pacific, with the west proving a real paradise for these nectar seeking insects which only reached California in the mid 19th century. Mostly they were wild swarms which made their nests in forests, and occasionally in abandoned houses.

Honey Hunters

If bees were not domesticated, it was a profitable pastime to hunt for honey, stealing it from bees' nests which might be found in hollow trees or rocky crevices.

The nests were tracked down by following the foragers. Usually this was achieved by luring them with some bait, such as a specially placed honeycomb which would allow the hunters to observe the returning beeline back to the nest. In some instances they would employ the services of birds with an inclination for raiding honey.

If the nest was in a tree the honeycombs would be removed, if necessary felling the tree to get to them. Alternatively the honey hunter would climb the tree, which goes some way to explain why the Russian name for beekeeper translates as 'tree climber'. In some countries those finding a honey tree would carve their mark to make it clear that they had found it. There could be severe punishment if another honey hunter stole from it – no less than 20 lashes and a fine in medieval Germany!

Wild honey nests were also hunted out and robbed in the most inhospitable places such as vertical cliffs, as demonstrated by that prehistoric cave painting at Cuevas de la Arana. This practice still continues in India, Nepal and other parts of the Far East. Compared to modern beekeeping it appears incredibly dangerous, with the honey hunter using long ropes to climb, a primitive back pack to hold the honey combs, and of course totally unprotected from the bees which even if stingless could still cause a lot of problems for someone dangling on a rope.

Folklore makes much of the pioneering spirit of the honey hunter, nowhere more so than in the USA. 'The Lost Honey Mines in Texas' tells several extraordinary stories. In one an old hunter relates how a man climbed up to a great cave above the Blaco River. On arriving, he was completely covered by thousands of bees and only saved from being stung to death by his heavy clothing. He drove the bees from his eyes for long enough to glimpse a solid wall of honeycombs. Later he returned with his companion, and with the aid of smoke and torches they were able to enter the cave. They approached the wonderful mass of honeycombs, but found the way was barred by a writhing mass of rattlesnakes. They retreated in terror. . . .

It beats putting on a clearing board for your honey gathering.

Early Hives

The ancient Greeks and Egyptians used hives made of pottery, shaped like giant thimbles and about the size of a modern skep. These would have been laid lengthways with entrance covers, probably stored in the walls of houses. Similar hives can still be seen in use in some countries.

The beekeepers of ancient Rome used all kinds of hives mentioned in their literature – log hives; hives made from cork bark; hives made from wooden boards; woven wicker hives; dung hives; earthenware hives; brick hives; and the aforementioned 'transparent stone' hives.

All these ancient hives were horizontal hives, and this type of design is still in use in many corners of the world today. They are usually made from hollow trees or logs, and can be found throughout primitive areas of Africa, South America and the Far East. Woven hives are also still frequently used in these areas, often with an outer coating of mud and dung. Apart from being used lengthways, these are not dissimilar from skeps.

In Europe skeps were the predominant type of hive from the Middle Ages onwards. The remains of a wicker skep found in the North Sea have been dated at AD 0–200, while a 12th century coiled straw skep of the type used today was excavated at York in 1980.

The straw skep probably evolved in Germany before spreading to the rest of Europe. Its one drawback was that the bees were generally killed to get at the honey! Those in the heaviest (the most honey) and lightest skeps (unlikely to survive the winter) were drowned at the end of the season, while the in-between skeps were allowed to survive. The combs could then be removed, and the empty skeps filled with swarms the following year.

Samuel Pepys

11 December 1663

'I to the coffee house. Mr Harrington told us how they do get so much honey. They make hallow a great Firr tree, leaving only a small slitt down straight in one place; and this they close up again; only leave a little hole and there the Bees go in and fill the bodies of these trees as full of wax and honey as they can hold; and the inhabitants at their times go and open that slit; and take what they please, without killing the bees, and so let them live there still and make more.'

An alternative method of 'driving the bees' into a different skep could be used if there was no brood, saving the colony but calling for advanced technique from the beekeeper. Demonstrations of driving bees are well worth seeing, with a rhythmic thumping used to persuade the bees that it's worth moving to the peace of a new home, despite abandoning their stores.

Another disadvantage of the skep is that it will obviously rot if exposed to too much bad weather. For this reason skeps were given shelter, such as 'bee boles' which were holes in walls used in Britain and Ireland. These were extensively used in the 17th and 18th century, and can still be seen in grand country houses – for instance the garden wall at Packwood House in Warwickshire (National Trust) has 30 along part of its length. In some cases the skeps would

be protected by special purpose-built 'bee houses'. These might be of quite grand design, and in the 19th century were adapted to house the new wooden hive designs.

Modern Forerunners

Wooden hives with removable honey chambers started to appear in Europe in the mid 18th century, and underwent a period of rapid development. If you went back a century or so you could probably see all sorts of splendid looking hives, but the real breakthrough came when the Reverend L. L. Langstroth invented the movable frame hive in the USA in 1851.

The concept was introduced to England by 1862 from where it spread rapidly throughout western Europe. Virtually all of today's hives are now similar in principal to that invented by the Rev Langstroth – WBC, modern Langstroth, and the 'National Hive Design' which was selected from current single wall hives as Britain's recommended hive by the Ministry of Agriculture and Fisheries in 1935.

Beekeeping Circa 1937

A 1937 beekeeping brochure.
W. Gale Bees & Hives shows that the main change in the last 50 years has been price! A complete WBC with brood chamber, two supers and frames is listed at 53 shillings; while a complete beginners' outfit – 'the best possible for those commencing beekeeping' – is £7 12s 6d including WBC, smoker, veil, feeder, queen excluder, clearing board, quilts ('to keep the bees warm'), a full complement of Italian or native bees, and a book of instruction.

With the exception of clothing which has improved greatly, the equipment appears to be little different from that which is available today. There is more plastic and the names have changed as have some of the designs – you don't see the splendid Holborn Hive or Cottagers Hive now – but the frames, foundation, smokers, sections, etc all look very similar. Perhaps it's time for some new designs?

Right: A selection of beekeeping equipment available over 60 years ago.

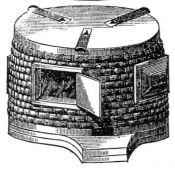

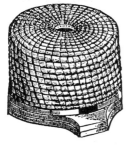

BOX FOR TRAVELLING BEES, No. 12f.

In this box with Combs secured with Rakes bees will travel safely for an unlimited distance. Bees brought from Cyprus in 1882 by me were conveyed in similar boxes.

Price **5/-**

Nos. 3, 4, 5, 5b, 6, 6b, 8, 8b, 9, 9b, and 11, on legs, **1/6** each, extra.

Painting any of the above Hives with 3 coats best paint, any color, **3/-** each. Graining and Varnishing or clear Varnishing any of the above Hives to order.

THE RUSTIC HIVE.—No. 12e.

A Novelty, being a Bar-frame Hive, constructed in the hollow trunk of a tree. Forms a pretty ornament for a lawn,

£3.

The number of these is limited, as the hollow trunks cannot always be procured some fine specimens are now on hand.

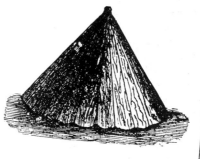

Manufacturer of Bee Keeping Appliances.

Appendices

International Bee Research Association

What is IBRA?
IBRA is an educational and scientific charitable trust, funded annually by subscriptions from its Members and by income from the sale of publications and other services.

What does IBRA do?
IBRA provides the world's most comprehensive information and advisory service on all aspects of bees and beekeeping – thereby helping to promote bee-keeping and bee research worldwide.

Why join IBRA?
★ To obtain comprehensive scientific and practical information and advice about bees and beekeeping.
★ To benefit from vast international collective agricultural expertise.
★ To support international agricultural research.
★ To become aware of the importance of bees in the environment.
★ To help underdeveloped countries by supporting IBRA's work in beekeeping aid projects.

International Bee Research Association and the usefulness of bees
From very early times, man has felt a special affinity with bees, first as the producers of honey, regarded as a gift from heaven, and later as providing an example of a well ordered society which seemed to reflect the society he himself sought. As a consequence, there has been much scientific enquiry about honeybees, and as they display a wide range of behavioural activities, are easily kept, and forage for their own food, they are an ideal experimental animal and have been used for many investigations: physiological, biochemical, behavioural, ecological and so on. Honeybees are most certainly the most studied social insect, possibly the most studied of all insects.

Background to the Association
The Bee Research Association (BRA) was founded in 1949 to advance research on bees and beekeeping. From the outset, the BRA set no national barriers to the scope of its work, and actively welcomed cooperation from any country; so in 1976 its name was changed to the International Bee Research Association (IBRA).

Dr Eva Crane was appointed as the first Director (a post she occupied until her retirement in 1983). She had both scientific qualifications and a great enthusiasm; added to this, a quite remarkable organizing ability and the gift of imparting her knowledge and enthusiasm to others. In the early days, Dr Crane's own home, first in Hull, Yorkshire, and then at Gerrards Cross, Buckinghamshire, was the headquarters of the Association, then as it grew there was a need for larger, permanent headquarters and in 1966 the Association moved to Hill House, Buckinghamshire, where it remained until 1986. In the economic climate then, with worldwide difficulties in obtaining grants and contracts, any short-term financial problems became potentially more serious. In addition, the cost of repairs to Hill House had become a burden which the Association did not have the resources to meet; so in October 1986 Hill House was sold in order to create the capital for a firmer financial base, and the headquarters were moved to Cardiff, where property values were much lower.

The current list of well respected bee scientists and beekeepers actively associated with IBRA is very comprehensive. IBRA's President is Dr M S Swaminathan (India); Dr Crane is Honorary Life President, and the Vice-Presidents include Brother Adam (UK), Professor G F Townsend (Canada), and Sir Vincent Wiggle-sworth (UK). Council Members include Professor J B Free (UK), Professor N Koeniger (GFR) and Dr G Vorwohl (GFR) and, until their recent retirements Dr J Louveaux (France) and Dr A Maurizio (Switzerland).

Aims and Objectives
IBRA is an educational and scientific charitable trust. The Association's work is funded annually by subscriptions from its Members (including institutions and other organizations), grants, donations and income from the sale of publications and other services.

One of IBRA's most important functions is to act as an international clearing house for scientific and technical information about honeybees, and other bees – their biology and role in the environment, and their products. This is achieved through its journals and other publications, which can be purchased or borrowed from the Association's library.

Membership
IBRA has Members and subscribers in over 100 countries, who recognize the Association as a primary source of information for scientist and beekeeper alike, and as the centre through which connection can be made with any aspect of bee research and the people interested in it. Scientific meetings and exhibitions are arranged from time to time in different countries, and IBRA's Regional Representatives provide links with Members. Members benefit from access to a vast international expertise, and by supporting IBRA's work they are helping international agricultural research especially in developing countries.

Library Services
At the Cardiff headquarters, IBRA Members have access to one of the most comprehensive agricultural lending libraries in the world. The Library receives current journals and publi-

cations from every part of the globe and all Members have access to this huge source of literature.

The main IBRA Library (named the Eva Crane IBRA Library in 1987) contains many thousands of books, reprints and periodicals. It holds many English translations of important publications, unpublished documents such as theses, manuscripts and important collections of letters. Members also have access to extensive subject and author indexes. Literature searches are conducted on request from Members and non-Members (these facilities are normally chargeable). Members may borrow books in person, or by post, and may obtain photocopies of publications. Branch libraries have been established in Columbia, Kenya, India and Japan.

IBRA Journals

Bee World is the official journal of the Association and is sent to all Members four times a year. Since its inception in 1919, *Bee World* has provided the international beekeeping community with news and authoritative articles and reviews on recent practical and scientific developments.

Agricultural Abstracts, published since 1950, gives an up-to-date summary of world literature and covers all aspects of research and technical developing concerning *Apis* and other bee species, including bee forage, hive products and pollination. The journal is produced using the CAB-I computerized system and thus forms part of a database which can be searched back to 1973.

The *Journal of Agricultural Research*, established in 1962, and now edited by Professor R Pickard, is a primary research journal publishing original research papers from all parts of the world. It has an International Editorial Advisory Board which includes: Dr Eva Crane (UK), Professor W Engels (GFR), Professor W E Kerr (Brazil), Professor C D Michener (USA) and Professor J Woyke (Poland).

In addition to journals, IBRA also publishes textbooks, reference books, bibliographies, newsletters, advisory leaflets and reports. As part of its information service, the Association operates an international mail-order book service.

Beekeeping in developing countries

An important aspect of IBRA's work is providing information to beekeepers and scientists in the tropics and subtropics. IBRA is recognized as a unique source of information on beekeeping programmes currently under way throughout the developing world. It maintains records of work in progress and completed projects in developing countries, and thus helps to ensure that resources are not wasted in repeating studies and other work.

The main work of IBRA's Information Officer for Tropical Agriculture, a post funded by the UK Overseas Development Administration, is to provide encouragement and advice on low-technology beekeeping for the rural poor in developing countries. Those involved in beekeeping in these countries are welcome to write for advice and information, and various free publications such as the *Newsletter*, leaflets and wall charts are available.

International Conferences

IBRA convenes the *International Conference on Agriculture in Tropical Climates*, which is hosted by governments around the world. Previous Conferences in the series have been held in London (UK), New Delhi (India) and Nairobi (Kenya). The fourth in the series was held in Cairo, Egypt, from 6–10 November 1988. These conferences bring together the knowledge and experience of participants from both developed and developing countries, thus stimulating contact and discussion between bee researchers, beekeepers, extension workers, specialists and representatives from international aid agencies concerned with agricultural and social development.

Recent Projects

The latest IBRA project to be completed, funded by the Commission of the European Communities, was an investigation to provide guidelines for the future development of beekeeping in Bangladesh. The objectives of this study were to survey nectar und pollen producing plants and to determine the beekeeping potential in rice-growing areas of Bangladesh. The project report concluded that although most of the species of plants that were important as bee forage and sources of honey were found at all the experimental sites, other species, particularly ones that provide forage during dearth periods, should be planted.

The majority of small farms in Bangladesh yield a very small gross income each year and there are also many landless peasants who earn even less. The extra income derived from just one honeybee colony could increase their income by 10%.

The Pollination Directory for World Crops was published by IBRA in 1984. Pollination is an essential stage in the reproduction of flowering plants and for some important crop plants there are many complicated factors determining the requirements for successful pollination. Few comprehensive books have been published on crop pollination and it has not always been easy to find out whether or not any specific crop requires action to ensure optimal pollination and thus a high economic yield. This book, funded by the New Zealand Ministry of Foreign Affairs, is intended to help provide this information and lists over 400 crop plants. It is of value to crop growers, agronomists, horticulturalists and foresters; to beekeepers and agricultural officers; and especially to those concerned with increasing food production.

Other recent grant-aided projects culminated in the publication of *Directory of important world honey sources*, the *Bibliography of Tropical Agriculture* and *The impact of pest management on bees and pollination*.

For information about membership contact:

International Bee Research Association 18 North Road, Cardiff CF1 3DY, U.K. Telephone: (0222) 372409/372450

National Honey Show

The National Honey Show has been an annual event since 1931. In recent years it has been held over three days at the Porchester Hall in London (W2) towards the end of October, giving plenty of time for the entries to be prepared.

It is by far the largest honey show in the UK with all manner of classes, and cash prizes and trophies for the winners. The Schedule Of Classes (1987 Show) which follows gives some idea of the scope available, and in addition there are nearly 90 more catering for NHS members and County Classes for Kent, Sussex, Essex, Middlesex and Buckinghamshire.

Schedule of Classes

Class No.

1. OPEN TO THE WORLD
   Twenty-four Jars of Honey. The exhibit may consist of honey of one, two, three or four kinds.

CLASS OPEN TO THE BRITISH COMMONWEALTH AND REPUBLIC OF IRELAND

2. **Three Jars of Honey** (Gift).—Any one colour or naturally crystallised.

CLASSES OPEN TO THE BRITISH ISLES INCLUDING THE REPUBLIC OF IRELAND

3. CRYSTAL PALACE
   Nine Jars and One Shallow Frame of Honey Suitable for extraction— The jars must be three each— Light, Medium and Naturally Crystallised.

4. DISPLAY
   Decorative Exhibit to Display Honey together with pure Moulded Beeswax or Mead or both.
   (In addition decorative material, coloured Beeswax and any size or shape jar may be used). Quantity of Honey staged to be at least 20 lbs. Size not to exceed 3ft 3in. square.

EXTRACTED HONEY

5. **Two Jars Light.—**
6. **Two Jars Medium.**
7. **Two Jars Dark.**
8. **Two Jars Chunk.**
9. **Two Jars Ling Heather.**
10. **Two Jars Soft Set.**
11. **Two Jars Naturally Crystallised (not soft set).**
12. **Twelve Jars Light, Medium, Dark, Naturally Crystallised or Soft Set, (Ling Heather excluded))** but all of the same; to be shown in 1lb. jars labelled as offered for sale. Beside the quality of the honey the attractiveness of the exhibit will be taken into consideration by the judges.
13. **3 lb. to 12 lb.**–to be exhibited in containers of any shape or type, labelled with the Exhibitor's own labels including name and address and in any form of display. To be judged equally for quality and sales appeal. The object of this class is to encourage originality and artistry in the presentation of honey for public sale.

COMB HONEY

14. **Two Sections Ling Heather.**
15. **Two Sections.** (Open only to Exhibitors who have not previously won this trophy).
16. **Two Sections free from Ling Heather.**
17. **One Comb, any source, any size.** To be suitable for extracting if other than Ling.
18. **Container of Cut Comb** (gross weight between 8 oz. and 12 oz.) Free from Ling Heather.
19. **Container of Cut Comb** (gross weight between 8 oz. and 12 oz.) Ling Heather.

BEESWAX

20. **One piece** (at least 1 lb. weight and at least 1 inch thick).
21. One Piece (minimum weight 12 oz.) prepared for commercial purposes. The block will be broken for judging, so perfect moulding is not required.
22. **Three Beeswax Candles. All made by Moulding, to be displayed erect.** One to be lit by the Judge.
23. **Three Beeswax Candles.** All to be made by any method other than by moulding to be displayed erect. One to be lit by Judge.
24. **Sheet of Foundation produced by exhibitor.**—Any method of production may be used. Size to be B.S. Deep or larger mounted in a frame and wired.

MEAD

25. **Mead, Dry.** (one bottle).
26. **Mead, Sweet.** (one bottle)
27. **Metheglin or Melomel, etc.**

MISCELLANEOUS CLASSES

28. **Any Interesting or Instructive Exhibit directly related to bees or beekeeping not including live bees**— Exhibits not previously awarded a prize at the National Honey Show (open to individuals only).
29. **A display of home-produced products containing honey and/or beeswax,** each item labelled for information.
   All displays of minimum 6, and maximum 10 items containing honey and/or beeswax. All containers will be opened for judging.
   This is a good class for the encouragement of home craft and artistry; **The display should include** items in which honey and/or beeswax plays an important part. Such as: Chutney Pickles, Honey Vinegar, Conserves, Preserves, Portions, Lotions, Cough cures, Cosmetics, Polishes, Models or moulds, Cakes, Crunchies, Cookies.
30. **An Attractive Model displaying Honey for Sale.**
31. **A practical Invention by the Exhibitor directly related to bees or beekeeping (No live bees).**—Not previously awarded a prize at the N.H.S. Awards of Merit will be given.
32. **Bee Skep** (Made by exhibitor) Unused.—Inside diameter between 12ins. and 18ins. Height greater than diameter.

PHOTOGRAPHIC CLASSES

All exhibits must have been taken—but not necessarily processed—by the entrant. The subject should be connected with beekeeping. Any number of entries may be made in each class. Each exhibit, properly mounted, shall be submitted in accordance with instructions issued by the Entries Secretary at the time of issuing labels. Transparencies or prints previously entered at N.H.S., whether prize winners or not, may not be entered again in the same class.

33a. Colour Transparency (Not close-up).—One only slide per entry— Any size film which fits into a 2in. by 2in. (5cm. × 5cm.) standard mount may be used.

33b. Colour Transparency (Close-up or macro).—Only one slide per entry—Any size film which fits into a 2in × 2in (5cm × 5cm) standard mount.

34. Mounted Colour-print photo—Only one print per entry—Minimum size of print 6in × 4in. (15cm. × 10cm.) mounted by the exhibitor on plain, tinted or coloured, cards 10in × 8n. (25.4cm. × 20.3cm.). Title to be shown on front of mount.

35. Mounted Black and White Photograph—Only one print per entry. Minimum size of print 6in × 4in (15cm × 10cm), mounted by the exhibitor on a white card 10in × 8n (25.4cm × 20.3cm). Title to be on front.

GIFT CLASSES Nos. 36–48
HONEY

36. **Two Jars Light.**
37. **Two Jars Medium.**
38. **Two Jars Dark.**
39. **Two Jars Ling Heather.**
40. **Two Jars Naturally Crystallised.**
41. **Two Jars Soft Set.**
42. **One Section.**
43. **Container of Cut Comb. Gross weight between 7 and 9 oz.— Labelled as for sale.**
44. **Six ½ lb. Jars Dark or Ling Heather.**

BEESWAX

45. **Six 1 oz. Blocks prepared for sale.**

CONFECTIONERY

46. **Honey Fruit Cake made as follows**—Recipe:—8ozs. self-raising flour, 6 oz. honey, 4 oz. butter (or marg, or mixed), 6 oz. sultanas, 2 eggs and pinch of salt. Cream butter and honey. Beat eggs well and add them alternately with sifted flour and salt. (Save a little flour to add with the sultanas. Beat all well and lightly. A little milk can be added if necessary. Turn in sultanas and stir well. Bake in a well-buttered circular tin, 6½ins. to 7½ins. diameter, for one and a quarter hours in a moderate oven.

47. **Nine Small Honey Cakes** not in paper cases. Total weight to be between 9 and 12ozs. Recipe to be submitted.

48. **Honey Sweetmeats. 1 lb box (Gross weight).**
NOTE: Box used must not bear any trade mark or reference to branded goods.

SCHOOL CLASSES

49. **For the Best Two Jars of Clear or Naturally Crystallised Honey** produced by a School Apiary. Entry forms must be accompanied by a list of the signatures of the children taking part in the apiary, and be made in the name of the school and signed by the Instructor.

50. **"An illustrated Beekeeping Note Book for the Active Season"** kept and entered by individual children under the age of 16 at the date of the show.

51. **Composite Class for Schools.** Three jars of extracted Honey, any one colour; One comb of Honey, any size, suitable for extracting or One Section or One piece of Cut Comb, and One Piece of Beeswax, weight between 7 and 9 ounces.

52. **An exhibition by Schools of School Beecraft** showing the educational value of beekeeping in schools.

MISS E.E. AVEY BEM, NDB. MEMORIAL CLASS

52a. Open to exhibitors aged under 25 years. One jar clear honey PLUS one jar naturally crystallised or soft set honey.

Publications

The National Honey Show publish in booklet form the following reprints of Feature Articles which have appeared in past Schedules:

No. 1—*Preparation of Liquid Honey*, by Cecil C. Tonsley.
No. 2—*Mead and Meadmaking*, by S. W. Andrews.
No. 3—*Wax for Show*, by F. Padmore.
No. 4—*Granulated or Crystallized Honey*, by C. C. Tonsley.
No. 5—*Production and Exhibitions of Comb Honey*, by W. S. Robson.
No. 6—*Management, Production and Exhibition of Heather Honey*, by George Vickery.
No. 7—*Judging Honey – in the Jar*, by C. C. Tonsley.
No. 8—*The Study of Pollen*, by Rex Sawyer.

A History of the Show is also published at a cost of £1 net post free.

Available from:
National Honey Show Publications
1 Baldric Road, Folkstone, Kent CT20 2NR

NHS Membership benefits

Free admission to the Show at all times.
No registration fee for exhibiting.
Advance copy of the Schedule and lecture programme free.
The right to exhibit in the special classes for members.
The right to speak and vote at the Annual Meeting held during the Show. Other members of family can become members for an extra small amount per person.
Free admission to the B.B.K.A. convention lectures which are held in conjunction with the Show.

Folbex VA versus Varroa

By Christiane Muschter/Ciba Geigy
Reproduced by kind permission of
CIBA Geigy

A. C. Oudemans, a Dutch zoologist, was the first to describe the varroa mite, in 1904. Mites belong to the arachnids, a class of arthropods that also includes spiders. Oudemans could hardly have foreseen that this species would one day turn out to be the beekeeper's most feared parasite and even the subject of newspaper headlines. He carried out his studies on Java, where varroa lives in a more or less balanced relationship with its host, the Asian bee. So the mite, although identified as a bee parasite, did not figure in the literature as a bugaboo of beekeepers and apiologists (people specialized in the study of bees). Not that is, until 50 years later, when reports of varroa-caused devastation of bee colonies in the Far East began appearing.

Tenacious adversary

What had happened? The mite, originally parasitic on *Apis cerana*, the Asian bee, on a line east of the Urals and Afghanistan, had switched its attentions to *Apis mellifera*, the European honey bee, which had been introduced into Asia on account of its greater productivity. Unlike its Asian cousins, however, the European bee has no natural defences against the parasite.

The adult female varroa, measuring 1.1–1.2 mm long and 1.5–1.65 mm broad, is ten times larger than the acarine mite. It is thus big enough to be seen with the naked eye, though it can readily be mistaken for the bee louse. Like all arachnids it has four pairs of legs, not always easy to spot because they are largely concealed by the light to dark reddish-brown dorsal shell. The yellowish-white male is much smaller and doomed to live but briefly: after mating it expires.

On its underside the mite has innumerable hairs. Like those of the bee they are pinnated, i.e. feathery, enabling the parasite to cling to the host. Varroa prefers to fasten onto that part of the bee where the epidermis is accessible and thin, particularly the abdomen, and where it can elude both the bee's cleaning activity and visual detection by the beekeeper. It bites the host and sucks the haemolymph – so-called bee's blood. Bacteria penetrating the open wounds can then cause infections that shorten the life-span of the bee.

Peril to the colony

The varroa mite parasitizes bees of any age and sex. It inflicts the greatest damage not on adult bees, however, which usually serve it as intermediate host and means of transportation, but on the larvae. The mites insinuate themselves into the brood cells shortly before the worker bees cap them with a protective covering of wax. There, inside the hive, the uninvited guest multiplies, laying its eggs soon after the bee larva has spun its silken cocoon. After 8–10 days the female mites and after 6–7 days the males are fully developed. Mating takes place while they are still in the sealed-off brood cell, following which the males, now no longer able to take food, die.

The mother mite and her offspring feed on the haemolymph of the bee larva, enfeebling the emergent worker or drone. The severer the mite infestation, the greater the damage: the young bees' abdomen is shortened and their wings or legs misshapen. Upon emerging at the same time as the new brood the mother mite and her offspring either attack them or attach themselves to adult bees.

The reduced vitality and decreased numerical strength of the young bee generation impair the colony so seriously that it usually dies off three years after the mite invasion. Investigations have shown that the bee larva's haemolymph plays a particularly important role in the development of the female varroa: only after having ingested this special food is she able to lay eggs. If only the haemolymph of adult bees is available, then she can survive but not reproduce. The presumption is that a juvenile hormone – a substance which programmes the maturation process – contained in the bee larva acts to stimulate her ovaries.

So well has the varroa mite adapted to its host that it even lives exactly as long as the bee: 2–3 weeks in summer, 2–4 months in winter.

Prone to the drone

Another noteworthy fact is that, while varroa mites attack all brood cells, including those of the queen, they show a preference for the drones. However, the mite is less selective in its behaviour towards the larvae of the European bee than towards those of the Asian bee, where it multiplies almost exclusively in the drone larvae – meaning that it can reproduce only during the brief period when the male brood is present. This, naturally, limits both the extent to which it can multiply and the damage it does, since a beehive can easily withstand the loss of a few drones. And that also explains why varroa does not inflict major losses in the Asian bees.

The parasite's development depends on the season. Reproduction begins in the spring, when the queen bee starts to lay eggs, with the varroa population-building up to a peak in the autumn. During the very nearly sterile winter period reproduction is scarcely possible. And although many of the mites die, most of them weather it out on the adult bees, which cluster together for warmth.

Spreading plague

As early as 1975 the magazine *Bee World* warned that, in the wake of the European bee's exportation to numerous countries and its assimilation abroad, varroa could become a world-wide threat. The prediction proved only too accurate. It was soon confirmed that the adaptable mite had managed to swap hosts and was widespread in several countries. During the 1970s it caused severe losses in bee populations in almost every part of the Balkans. In Bulgaria alone, around 200,000 colonies perished within the space of three years. A Soviet expert described the losses due to varroa disease in his country as being greater by orders of magnitude than those caused by all other diseases of bees taken together.

This was alarming news indeed, because institutions and private beekeepers in Germany had more than once obtained queen bees from the Soviet Union and Southeast Europe. Moreover, whole colonies had been transported over great distances for the pollinating of various areas rich in flowering plants.

The further spread of the varroa epidemic in Europe had to be expected. Even so, it came as a shock when the mites were discovered for the first time in West Europe in the Taunus region of the Federal Republic of Germany in February 1977. At that point, actually, the parasite was already more widespread than suspected. And soon afterwards centres of infection were reported in the states of Baden-Württemberg, Rhine-Palatinate, Bavaria and Lower Saxony as well as Berlin. An official prohibition on the transport of affected colonies failed to stay the plague.

How to stop it?

Other countries fared no better. Within one year the mite vaulted the considerable distance from Macedonia to Istria, and by 1981 there were no bees left in Dalmatia. The following spring the parasite turned up in Italy. It was brought into South America from Japan, where the European bee is also kept, and imported into North Africa from Romania. Today varroa can be found on every continent save Australia. Thanks to strict import controls, and heavy fines imposed on violators, Switzerland (*and the UK*) has also been spared – so far.

Immediately the mite had been spotted in Germany, a search for ways and means of containing it set in. Biological control methods and even the destruction by burning of entire colonies brought no success. Substances that had been tried out in the countries where the epidemic had appeared years previously were only moderately effec-

tive. The intensive quest for a defensive weapon which would be destructive to the mites but which the bees could tolerate now became a race against time.

One key site of this effort was the Apiology Institute in Oberursel, Hesse. Nor could Give-Geigy ignore the challenge: *Folbex*, a Geigy product introduced in 1952, had proved effective against the less problematic acarine mite. But against varroa its effect fell short of what was needed. In April 1980 the Biotechnical Branch of the Agricultural Division in Basel therefore decided to push ahead with the development of a new product – a medicament designed expressly to combat the parasite.

A bee drug takes shape

Charged with heading the project were René Fatton and Wolfgang Schmid. 'After analyzing the requirements listed by the authorities and the available resources,' Dr Fatton related, 'We drew up a detailed plan for the development of the new product. Thanks to first-rate teamwork by everyone who participated, we were largely able to keep to it.'

In a sense, that is a singular understatement. The period which elapsed from activation of the project in the spring of 1980 until registration of the product, *Folbex VA*, just two years later was remarkably brief. One important expediting factor was the choice of fumigant strips as the form of application – a decision based on experience with the first Folbex formulation. This had shown that fumigation was an efficient means of distributing the active substance in the beehive.

The first prototypes of the sought-after product were already available for testing in May 1980. 'We struck it lucky in being able to work closely with Dr Wolfgant Ritter, the bee specialist who had just taken over the Department of Apiology at the Institute of Veterinary Hygiene in Freiburg im Breisgau,' declares Dr Fatton. (Freiburg i.B. is less than an hour down the Rhine from Basel.) Dr Ritter had previously made

his reputation at the Apiology Institute in Oberursel as an apiarist, author, and member of international working groups set up to combat varroa disease. As early as June 1980 he determined that the new Folbex formulation was surprisingly well tolerated by bees. That same month, during the foraging period, the first residue test was carried out. In July, while in Tunisia for the UN Food and Agriculture Organization, Dr Ritter tested the Ciba-Geigy medicament together with other products in the field. 'The results,' says Dr Fatton, 'exceeded our expectations.'

These encouraging preliminaries spurred the development process, and further trials confirmed the good initial results. They also showed that Folbex VA was excellently suited for the diagnosis of varroa disease – an important property, since early recognition makes it easier to nip an infestation in the bud. Soon after the discovery, the authorities found an occasion to put it to good use. In the South Baden town of Waldshut, just opposite the Swiss bank of the Rhine, varroa mites were found in a beehive in 1981. The Swiss responded by combing a front ten kilometres wide along the river from Basel up to Schaffhausen for other evidences of the parasite. To the relief of the officials and beekeepers, none turned up.

Berlin and Berne approve

'It was envisaged that the new compound should also be effective against the acarine mite, which is widespread in our part of the world,' explains Wolfgang Schmid. 'With that in mind we undertook to have field trials carried out by the Institute of Veterinary Hygiene in Freiburg together with the Veterinary College of Toulouse, France.'

In Switzerland, similar investigations were performed by Dr Hans Wille, head of the Bee Section in the Federal Dairy Farming Research Institute, Liebefeld. These tests, as well as extensive residue trials, showed favourable results throughout. 'And so,' Wolfgang Schmid continues, 'after compiling and sifting all of the data, inclusive of a summary expert opinion covering all the various sectors involved, we were able in collaboration with our Group company in West Germany to apply for official approval. On July 27, 1981, we submitted the documentation to the Federal Office of Health in Berlin. It weighed more than 30 kilograms!'

Two months later registration was also applied for in Switzerland, and in February 1982 came the good news: Folbex VA was admitted as the first officially approved drug for bees in both countries. While the new product does have the name 'Folbex' in common with its predecessor, the suffix 'VA' indicates that it is effective against both varroa and acarine disease.

The medicament was already being applied in February 1982 – just in time for its use as a diagnostic aid in detecting newly infested areas in Germany and for controlling acarine mites in Switzerland.

A briskly paced success story with a taste of honey, then. But not one with a facile happy end, even with Folbex VA available. Based on discussions with experts from the Apiology Institute in Oberursel and with Dr Ritter in Freiburg, the status and the prospects can be summarized as follows:

- In the absence of control measures the varroa mite poses a mortal threat to European bee colonies, and in West Germany, at least, it has been found to be a great deal more dangerous than was at first assumed. In South Hesse and North Baden alone, the parasite destroyed more than 2000 colonies in the fall of 1982.
- Complete eradication of varroa disease will not be possible. In other words, beekeepers are going to have to learn to live with it.
- All of the biological control methods tried to date, such as removal of the infested drone brood, have failed to bring the hoped-for success. Apiarists will therefore have to depend on chemotherapy to keep the varroa mite in check.
- In the continuing fight against this tenacious parasite. Folbex VA is an advanced defensive weapon.

University College Cardiff Bee Research Association

The Bee Research Unit at Cardiff is the largest centre of bee research in the UK and has attracted many donations from beekeepers and their associations since its foundation in 1977. A new society, the University College Cardiff Bee Research Association, has now been established to give continuous financial support to the Bee Research Unit through its membership subscription fees.

The society elects its own principal officers, presents its own independently audited accounts to any AGM, and any member is entitled to stand for election to its executive committee. The headquarters of the new association are housed alongside the Bee Research Unit's laboratories in the centre of Cardiff.

Membership is available to any persons wishing to promote the future development of the Bee Research Unit at Cardiff in all its many activities.

Research

At Cardiff, students can investigate any aspect of honeybee biology and any practical problem in apiculture at either Final Year B.Sc., Diploma, M.Sc. or Ph.D. level. They have access to the Department of Zoology's three electron microscopes, analytical laboratories, computing services and technical expertise in most branches of the biological sciences.

The Bee Research Unit is well equipped with data-gathering electronics and the full range of beekeeping equipment needed to manage a small commercial apiary. Field-work and experiments on bees under glass are undertaken at the University's Cleppa Park Field Station.

The following research subjects are currently being studied at Cardiff:

Acarapis
ageing
biosensors
bumblebee domestication
computerised neuroanatomy
electrophysiology
instrumental insemination
Nosema
pheromones
pollen indentification
pollination behaviour
queen-propagation
sensilla
social behaviour
tropical apiculture

Diploma in Apiculture

In addition to its research activities, the Bee Research Unit provides lectures and practical courses from October to June each academic year for the University's international Diploma in Apiculture:

TERM ONE – BIOLOGY
Evolution
Anatomy
Cytology
Physiology
Behaviour
Pathology
Ecology
Beekeeping
Honey

TERM TWO – MANAGEMENT
Production – honey
              wax
              pollen
              royal-jelly
Queen propagation
Bee breeding
Crop pollination
Finance
Marketing

TERM THREE – Completion of research project
To date, the Unit has trained students from Bangladesh, Colombia, England, Egypt, Iraq, Japan, Kenya, Mexico, Mozambique, Sudan, Tanzania, Turkey, Wales and Zambia. Through this international course, the Unit is able to make a practical contribution to the establishment of low cost, low technology apiculture in many developing countries.

UK Beekeeping and Bee Health Statistics 1987

The figures in the four tables which follow were provided by the Ministry of Agriculture, Fisheries and Food Horticulture Branch, and were the most up to date available at the time this book went to press in 1988. There is no official registration of beekeepers in the UK, and as not all apiaries are inspected each year these figures are necessarily only estimates (© Crown Copyright 1988).

Number of recorded beekeepers and colonies with analysis of beekeepers by number of colonies owned

	Number of Beekeepers with live Colonies					Total number of colonies	Number of colonies owned by beekeepers with 40 or over colonies
	owning 1 to 10 colonies	owning 11 to 39 colonies	owning 40 to 100 colonies	owning over 100 colonies	Total		
ENGLAND							
1985	30,610	2,741	285	114	33,750	179,461	41,554
1986	29,047	2,476	246	92	31,861	162,548	34,334
1987	28,183	2,238	249	92	30,762	150,551	33,767
WALES							
1985	2,380	266	35	12	2,693	15,455	4,324
1986	2,321	258	33	10	2,622	12,885	2,894
1987	2,325	185	23	9	2,542	10,039	2,458
ENGLAND AND WALES							
1985	32,990	3,007	320	126	36,443	194,916	45,878
1986	31,368	2,734	279	102	34,483	175,433	37,228
1987	30,508	2,423	272	101	33,304	160,590	36,225

*There is no official registration of beekeepers and as not all apiaries are inspected each year these figures are necessarily only estimates.

Inspections under the bee diseases control order 1982

	Number of colonies inspected			Number of diseased colonies destroyed or treated in the year			Number of diseased colonies destroyed or treated as percentage of colonies inspected	
	LIVING	DEAD	TOTAL	AFB	EFB	TOTAL	AFB	EFB
ENGLAND								
1985	59,584	7,589	67,173·	433	368	801	0.64	0.55
1986	45,294	17,623	62,917	291	121	412	0.46	0.19
1987	43,523	9,923	53,446	266	156	422	0.50	0.29
WALES								
1985	6,241	493	6,734	49	113	162	0.73	1.68
1986	4,616	3,140	7,756	25	15	40	0.32	0.19
1987	4,410	1,836	6,246	16	3	19	0.26	0.05
ENGLAND AND WALES								
1985	65,825	8,082	73,907	482	481	962	0.65	0.65
1986	49,910	20,763	70,673	316	136	452	0.45	0.19
1987	47,933	11,759	59,692	282	159	441	0.47	0.27

Results of laboratory examination of sample combs

Laboratory service provided by ADAS	Number of combs received	Laboratory Diagnosis		
		AFB	EFB	No Disease
Samples Submitted by Bees Officers				
1985	855	138	481	115
1986	347	89	136	119
1987	398	108	157	107
Samples Submitted by Beekeepers				
1985	43	8	1	23
1986	30	2	-	23
1987	17	1	2	14
Total (England & Wales)				
1985	898	146	482	138
1986	377	91	136	142
1987	415	109	159	121

Examination under the importation of bees order 1980 of queen bees imported into England in 1987

Country of origin	No of queens imported	No of consignments imported	Total samples* of worker bees examimed	Nosema	Other disease	No disease
New Zealand	2013	45	45	3	-	42
USA	2613	77	77	22	1†	54
Others	127	6	6	-	-	6
Total	4753	128	128	25	1†	102

* A sample may comprise all the workers from a consignment of queens.

† Acarine

The National Diploma in Beekeeping

Hon Secretary: R.E. Gove N.D.B., Westcott, Gerway Lane, Ottery St Mary, Devon EX11 1PW
Chairman: G. Hopkinson N.D.B., County Inspector, Staffordshire County Council, County Buildings, Tipping Street, Stafford ST16 2DH

The Examination Board for the National Diploma in Beekeeping was set up in 1954 to meet a need for a beekeeping qualification above the level of the highest certificate awarded by Beekeeping Associations and approaching degree level.

The need was manifest as an appropriate qualification for a County Beekeeping Lecturer or a specialist appointment requiring a high level of academic and practical ability in beekeeping. It is the highest beekeeping qualification recognised in the United Kingdom and internationally and those who obtain the Diploma can feel justly proud.

The Board consists of representatives from a wide range of organisations and from Government Departments and together form an impressive amalgam of expert knowledge in beekeeping and education. Although the National Beekeeping Associations are represented on the Board it is entirely independent of them.

Normally the highest certificate of one of the National Associations is a necessary criterion for eligibility to take the Examination for the Diploma, which is held in alternate years and extends over two days. The Examination consists of two written papers of three hours each and viva-voce, plus practical tests conducted by at least four Examiners appointed by the Board.

The Board also organises, in conjunction with the Hampshire College of Agriculture, an Advanced Beekeeping Course covering certain parts of the Syllabus which can be difficult to study at home. It lasts a working week and covers a wide field including laboratory techniques, bee disease recognition, the present day problems of honey packing plants and commercial beekeepers at their respective premises. The outside speakers are leaders in their fields.

For further details write, *enclosing s.a.e., to the Secretary.*

ADAS National beekeeping unit

Luddington Experimental Horticulture Station, Stratford-upon-Avon, CV37 9SJ Tel 0789 750601
The ADAS National Beekeeping Unit provides a statutory and advisory service to beekeepers in England and Wales. The Unit has about 100 colonies of bees which are used for investigational work and demonstrations. Current investigational work includes studies of the effects of pollen supplement feeding on the early spring development of bee colonies, the role of honey bees as pollinators of fruit crops and the effect of Nosema and Acarine diseases on infected untreated colonies.

Advice

Advice is given by correspondence on all aspects of apiary management and is supported by advisory leaflets, lectures and demonstrations, mainly to conferences and meetings organised by beekeeping associations, in many parts of England and Wales. Practical seminars on various aspects of beekeeping are also held at Luddington. The subject matter is varied but emphasis is presently being given to basic colony management for honey production, bee disease control, rearing of queen bees, the beekeeping potential of oilseed rape and varroa detection techniques.

Diagnostic service
The Unit provides a diagnostic service for beekeepers who suspect their bees are suffering from adult bee diseases. The service is for beekeepers where none is provided by a County Beekeeping Instructor.

Statutory work
The Unit's statutory responsibility mainly concerns implementing provisions of the Bee Diseases Control Order 1982 and the Importation of Bees Order 1980. Under the Bee Diseases Control Order 1982 beekeepers are obliged to notify the appropriate Agriculture Department if they suspect cases of American foul brood, European foul brood or Varroasis. Combs of brood suspected of being infected with foul brood are examined in the laboratory at Luddington. Most combs are sent in by Bees Officers who inspect beekeepers' colonies from April to September inclusive throughout England and Wales. Some combs are received direct from beekeepers. The Unit's staff also has a responsibility in the training of Bees Officers.

Under the Importation of Bees Order 1980 escort worker bees which accompany queen bees imported into England and Wales must be sent to Luddington where they are examined for mites and bee disease organisms.

Major Honey Plants

Blackberry
Rubus Fruticosus
Wild or cultivated, the blackberry flowers late in the summer (June–August), providing nectar and pollen at a time when our sources are scarce.

Buckwheat
Fagopyrum Sagittatum
A grain crop which is widely grown in the USA, USSR and parts of Europe. It produces a dark honey.

Charlock
Sinapis Arvensis
A mustard relation with bright yellow flowers that is regarded by farmers as a weed.

Dandelion
Taraxacum Officinale
A wild plant growing throughout the UK, much of Europe, and North America and New Zealand.

Clovers;
A most important honey producing plant in the UK and elsewhere throughout the world – in particular Canada and New Zealand. White clover is the leading variety.

Crimson Clover	*Trifolium Incarnatum*
Hop Clover	*Trifolium Campestre*
Red Clover	*Trifolium Pratense*
Swedish Clover	*Trifolium Hybridum*
Yellow Clover	*Trifolium Dubium*
White Clover	*Trifolium Repens*

Field Beans
Vicia Faba
Field beans are a major crop for honeybees, as are broad beans if available. Runner beans and French beans are not so well suited to their needs.

Fruit Blossom:
An early source of nectar and pollen, most useful for brood rearing. The apple and cherry are most productive for the honeybee; the fig and mulberry not at all so.

Apple	*Malus Pumila*
Pear	*Pyrus Communis*
Plum	*Prunus Domestica*
Cherry	*Prunus Cerasus*

Hawthorn
Crataegus Monogyna/Oxyacantha
A native tree or shrub with white flowers that appear in May. It can be a good source of nectar.

Heather (Ling)
Calluna Vulgaris
Used for specialist honey production in the UK, most of which is in Scotland and on high moorland. A migratory beekeeping technique is used.

Lime (Linden)
Tilia
A major honey producer in the UK, particularly in towns where limes are widely planted.

Mustard/Black
Brassica Nigra
Mustard/White
Sinapis Alba
Commercial mustard crops are good nectar producers for honeybees.

Oil Seed Rape
Brassica Napus
A unique plant producing its own major spring honey flow.

Sainfoin
Onobrychis Viciifolia
A fodder plant grown throughout Europe which is a good nectar producer.

Sycamore
Acer Pseudoplatanus
Widespread throughout the UK; flowers towards the end of May and is a good source of nectar.

Willow Herb
Epilobium Angustifolium
A wild bee plant which can be a good source of nectar in the UK: also widespread in Canada.

Beekeeping Journals

American Bee Journal
The oldest magazine in the English language. Professional and scientific news; worldwide crop and market page.
American Bee Journal, Hamilton, Ill 62341, USA.

Apiacta
International beekeeping magazine published in English by the International Federation of Beekeepers' Associations; facts and articles from the world of beekeeping.
Apimondia, Corso Vitt, Emanuelle 101, 00186, Roma, Italy.

Australasian Beekeeping
Senior beekeeping journal of the Southern Hemisphere. Complete coverage of all beekeeping topics in one of the world's largest beekeeping countries. Published monthly.
Pender Beekeeping Supplies, PMB, 19 Gardiner St, Rutherford, NSW 2320, Australia.

Bee Craft
Official journal of the British Beekeepers' Association, published monthly.
Bee Craft Secretary, 15 West Way, Copthorne Bank, Crawley, Sussex RH10 3QS.

Beekeepers News
Quarterly publication produced by Thorne Beehives.
E. H. Thorne (Beehives) Ltd, Beehive Works, Wragby, Lincoln LN3 5LA.

Beekeepers Quarterly
Large format news, views and feature magazine.
Beekeepers Quarterly, Scout Bottom Farm, Mytholmroyd, W. Yorkshire HX7 5JS.

Beekeeping
A West Country bee journal; 10 issues per annum.
The Editor, Clifford Cottage, 42a Clifford St, Chudleigh, Devon TQ13 OLE

Bee World
International English language beekeeping quarterly, published by the International Bee Research Association.
IBRA, 18 North Road, Cardiff CF1 3DY.

British Bee Journal
Monthly news, views and reviews magazines.
British Bee Publications Ltd, 46 Queen Street, Geddington, Kettering, Northants NN14 1AZ.

Canadian Beekeeping
PO Box 128 Orono, Ontario, Canada.

Canadian Bee Journal
47 Black Knight Road, St Catherine's Ontario, Canada.

Gleanings In Bee Culture
Easy to read, new and better ideas for keeping bees. Published monthly.
PO Box 706, Medina, OH 44258–0707, USA.

Indian Bee Journal
The official organ of the All India Beekeepers Association. The only Indian journal, published all in English with articles and information of interest to beekeepers and bee scientists, particlarly concerning Apis Cerana Indica and other species of bees of the Indo Malayan region.
Indian Bee Journal, 1325 Sadashiy Peth, Poona 411030, India.

Irish Beekeeper
'An Beacaire' – published monthly.
Managing Editor, J. J. Doran, St Judes, Mooncoin, Waterford, Ireland.

The Australian Bee Journal
A magazine carrying interesting Australian news.
Willunga, RMB 4373, Moonambel 3478, Australia.

The New Zealand Beekeeper
Quarterly magazine published by the National Beekeepers' association of New Zealand. Practical beekeeping, latest research and featuce articles with large format and many illustrations.
NZ Beekeeper, Box 4048, Wellington, New Zealand.

The Scottish Beekeeper
Magazine of the Scottish Beekeepers' Association – international in appeal; Scottish in character.
D. Blair, 44 Dalhousie Road, Kilbarchan, Renfrewshire PA10 2AT.

The Speedy Bee
Regular magazine features.
PO Box 998, Jessup, CA 31545 USA.

South African Bee Journal
The leading bee journal in Africa; official organ of the Federation of Beekeepers' Associations published bi-monthly in English and Afrikaans.
The Chairman, PO Box 4488, Pretoria, 0001 South Africa.

Useful addresses

Agricultural Education Association
Secretary: Norfolk College of Agriculture, Easton, Norwich NR9 5DX. Tel: 0603 742105.

Bee Disease Insurance Ltd
Secretary: Pump Cottage, Weatheroak, Nr Alvechurch B48 7EQ.
Tel: 0564 822059.

Bee Farmers' Association
Secretary: Southfields, The Crossways, East Markham, Nr Newark NG22 0SG.
Tel: 0777 870968.

British Beekeepers' Association
National Beekeeping Centre, National Agricultural Centre, Stoneleigh, Kenilworth CV8 2LZ.
Tel: 0203 552404.

British Isles Bee Breeders Association
Secretary: 11 Thomson Drive, Codnor, Derbys DE5 9RU.

Central Association of Beekeepers
Secretary: Long Reach, Stockbury Valley, Sittingbourne ME9 7QP.

Council of the National Beekeeping Associations of the UK
Secretary: 12/27 Ethel Terrace, Edinburgh EH10 5NA.
Tel: 031 5332.

Devon Apicultural Research Group
The Secretary: 20 Parkhurst Road, Torquay, Devon TQ1 4EP.

Federation of Irish Beekeepers' Associations
The Secretary: 45 Waltham Terrace, Blackrock, Co Dublin, Eire.

Gloucester Honey Bee Improvement Group
The Secretary: Ryelands, Ross on Wye HR9 7TU.
Tel: 098985 257.

International Bee Research Association
18 North Road, Cardiff CF1 3DY.
Tel: 0222 372409.

MAFF Horticulture Division
Branch A, Great Westminster House, Horseferry Road, London W1.
Tel: 01 216 6717.

National Diploma in Beekeeping
The Secretary: Westcott, Gerway Lane, Ottery St Mary, Devon EX11 1PW.

National Honey Shows
The Secretary: Gander Barn, Southfields Road, Woldingham, Surrey CR3 7AP.
Tel: 088 385 3152.

Scottish Beekeepers' Association
The Secretary: 9 Glenhome Avenue, Dyce, Aberdeen AB2 OFF.

Ulster Beekeepers' Association
The Secretary: Moyola Lodge, Castledawson, Co Derry.
Tel: 0648 68224.

University College Cardiff Bee Research Association
Bee Research Unit, Department of Zoology, University College Cardiff CF1 1XL.
Tel: 0222 874312.

Welsh Beekeepers' Association
The Secretary: Woodlands, Llandrindod Wells, Powys, Wales.

Welsh Beekeeping Centre
Coleg Howell Harris (Brecon College of Further Education), Penlan, Brecon/Aberhonddu, Powys LD3 9SR.

Index